ALACTIC PROJECTION

Milky Way, drawn by M.T.Keskūla, Lund

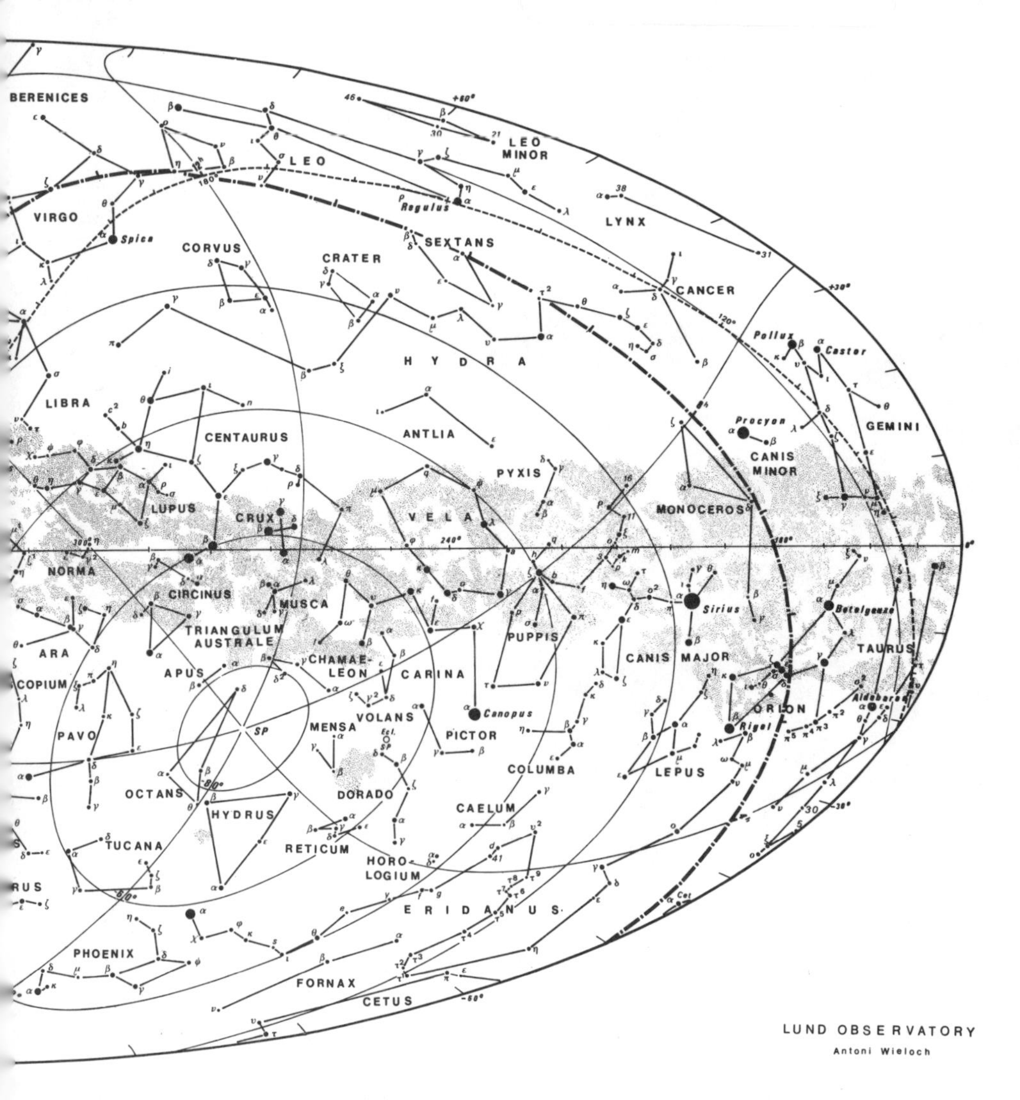

The Milky Way

The Milky Way

The Structure and Development of our Star System

Ludwig Kühn
Astronomical Institute of the Ruhr University,
Bochum, Federal Republic of Germany

JOHN WILEY & SONS
Chichester · New York · Brisbane · Toronto · Singapore

First published under the title Das Milchstrassensystem: Bauplan und Entwicklung unseres Sternsystems by Ludwig Kühn

Library of Congress Cataloging in Publication Data:

Kuhn, Ludwig.
The Milky Way.

Translation of: Das Milchstrassensystem.
Includes index.
1. Milky Way. I. Title.
QB857.7.K8313 523.1′13 82-2820

ISBN 0 471 10277 6 AACR2

British Library Cataloguing in Publication Data

Khun, Ludwig
The Milky Way.
1. Milky Way
I. Title. II. Das Milchstrassensystem. *English*
523.1′13 QBi57.7

ISBN 0 471 10277 6

Typeset by Activity, Salisbury, Wilts., and printed in the United States of America.

Contents

Preface

In spite of being one of the oldest sciences, astronomy is still in its infancy. New celestial objects constantly steal the limelight and new equipment and techniques have exposed new wavelength ranges, swinging the emphasis away from the descriptive and more towards an interpretive attitude. Its close links with physics always were and remain of central importance.

Until the nineteenth century, astronomers were chiefly concerned with the dynamics of the planetary system, but from then on interest switched to the physical properties of the stars and interstellar matter, using the recently discovered techniques of photography and spectroscopy. It was not long before the stars were recognized as members of a large system, but one which was not a unique phenomenon in the Universe.

Our galaxy, the Milky Way, is now a focal point for astronomical research. Highly sophisticated radioastronomy equipment has defined the spiral structure of our system over a wide range and parts of the Milky Way which were invisible until very recently can now be picked up in the infrared and radio range. The advent of large, fast computers has meant that the movement of thousands of stars and interstellar gases can be plotted, thus assisting research into the dynamics of complicated mass systems such as the Milky Way. Added to this, the density wave theory of spiral structure and the in-depth knowledge available on star development provided new incentives for research into the formation and development of the star system. The Milky Way is the best known star system mainly because we see it from the inside, whereas as far as the others are concerned, we are on the outside looking in.

A great many generalized books on astronomy give merely a short descriptive outline of the Milky Way; the objective of this book is to present it as a large, ever changing system, whilst keeping the subject as generalized as possible and using no highly specialized terminology, so as to appeal to a wide range of readers. Tables at the end of the book summarize the most important data on the Milky Way, the units used throughout and corresponding conversions.

Several colleagues have helped me to compile this book during the course of many discussions and by means of critical comment, but I would like to acknowledge in particular the assistance afforded by Professor Th. Schmidt-Kaler of Bochum.

Bochum, January 1978 *Ludwig Kühn*

I

Stars seen by the naked eye

Historical evidence shows that man has always studies the heavens attentively, whether it be to predict the weather from cloud formations or to read the time of day or the seasons from the position of the Sun. But the clear night sky in all its complexity and with all its variations has also been a constant source of fascination. The most striking object was the Moon with its constantly changing shape, and the relatively fast-moving planets, passing through a network of apparently uniformly rotating fixed stars. However, even this fixed star background changed during the course of a year and travellers reported seeing different stars and familiar star groups in different parts of the world. The seasonal change in sky patterns helped ancient seafarers to work out their bearings when land was no longer in sight. Apart from these practical aspects, the size and glitter of the heavens have always fascinated man. Various constellations in the firmament were taken to represent the shapes of gods and heroes and thus the whole sky acted as a backdrop for the ancient tales of religion and brave deeds. Those based on Greek mythology are the most familiar zodiac patterns remaining and indeed in use today. The southern hemisphere was not seen by the Greeks and was divided up into constellations during the great period of exploration and discovery, many being tagged with very 'modern' names such as Telescopium and Microscopium.

The whole sky is now divided up into 88 constellations: 26 in the northern sky, 12 in the zodiac signs and 50 in the southern sky. In addition to the prominent constellations there is one special phenomenon in the fixed star system which is seen by us mostly in summer and autumn, and that is the bright, laminar strip known as the Milky Way. It encircles the whole firmanent, beginning at the constellation Cassiopeia and running through Cepheus, Cygnus and Aquila, crossing the zodiac signs in Sagittarius and Scorpio, touching the southern hemisphere around the Southern Cross, crossing the zodiac again in Gemini and then via Auriga and Perseus back to Cassiopeia again. The strip is not uniform in width, but is rather irregular, bulges in places and seems to be split into two in the Cygnus constellation. As with so many of the configurations, this is also associated with Greek mythology. The story goes that when Zeus put his son Heracles, conceived and borne by a mortal, to the breast of his sleeping wife Hera, in order to ensure his immortality, the infant sucked so lustily that the goddess's milk spurted right up to heaven, forming the Milky Way. The Greek materialist Democrates put forward the theory that in reality the luminous celestial strip consisted of a mass of individual stars, but this was not confirmed until the time of Galileo and the invention of the telescope.

However, not only clearly visible stars were formed into constellations at that time, but all stars visible to the naked eye were counted and classified. The brightest stars were given the 'magnitude 1' rating and the least visible were categorized as 'magnitude 6'. The catalogue compiled by the Greek astronomer Hipparch, 190–126 BC, and published by Ptolemy in the Almagest, lists 1022 stars divided up into 21 constellations for the northern hemisphere, 15 for the southern hemisphere and 12 zodiac signs. Even though the stellar brightness classification was rather rough and ready, it is still used today, except that each category has now been much more closely defined and standardized so that every one of the six groups contains concise data. For example, the intensity ratio between two stars which are only one brightness magnitude apart is equal to 2.512. According to this, a star with a magnitude 2.5 times greater than another has an emission level 10 times lower.

We know that not all stars visible from the island of Rhodes, where Hipparch lived for many years, are listed in his catalogue. A much more comprehensive study was carried out by John Flamsteed, the first astronomer of the Royal Greenwich observatory, whose catalogue was the first based on telescopic studies, contained 2866 entries and appeared at the beginning of the eighteenth century. There are about 6000 stars which can be seen with the naked eye in the northern and southern hemispheres combined, and this number increases dramatically if those which are still only visible by telescope are included too. The Bonner Durchmusterung carried out by Friedrich Wilhelm Argelander and Emil Schönfeld in 1862 listed 324,198 stars up to magnitude 9 and even a few of magnitude 10, in the northern hemisphere. The figure is increased yet again by including all stars viewed through the largest available telescope. If photography is used, then a single plate representing a four-thousandth part of the sky shows about 150,000 stars up to and including magnitude 16. A large photographic survey of the heavens, reproducing all objects up to magnitude 21, reveals about one billion* stars. However, distribution is not uniform and there are particularly high concentrations of stars in the Milky Way. A large photographic survey carried out from Mount Palomar in California with a 1.2 m Schmidt mirror telescope reveals in a one square degree section of the Milky Way about 100,000 stars; a similar area outside the Milky Way, around the Pole, for example, contains about 1500 stars.

*1 billion = 1000 million = 10^9.

Figure I.1. This overall picture of the Milky Way gives a clear view of the concentration in the Milky Way plane or galactic equator. The network plotted represents the galactic co-ordinates. The brightness in the galactic plane is due to star concentration of bright young stars and H II areas in particular. The dark strips embedded in the bright patches are caused by absorption of interstellar dust. The Large and Small Magellanic Clouds are found in the bottom right-hand section at longitude 270° and 245°, respectively. The star system M 31 in the Andromeda constellation is seen to the left below the galactic equator at longitude 130°, latitude −20°. (Photograph: Lund Observatory)

Figure I.2. The photographs show one third of the Milky Way and were taken with a spherical mirror camera which has an extremely wide angle. The centre of the Milky Way in the Sagittarius constellation is clearly seen just to the left of centre of the bright Milky Way band and is particularly well reproduced in the red spectrum range. The light-absorbent dust complexes occur mostly to right and left of centre and form a much narrower strip in the centre of the Milky Way. The UV photograph clearly shows the shingle-type effect of the Milky Way. (a) UV spectrum (wavelength ~ 350 nm); (b) visual photograph (wavelength ~ 550 nm); (c) red spectrum (wavelength ~ 650 nm). (Photographs: W. Schlosser, Th. Schmidt-Kaler and W. Huenecke, Ruhr-University, Bochum)

Number of stars per square degree (galactic length l = 180°)

Limiting magnitude	Galactic width		
	b = 0°	b = 40°	b = 80°
6m	0.12	0.04	0.03
12m	75.9	17.8	11.0
18m	10,000	832	550

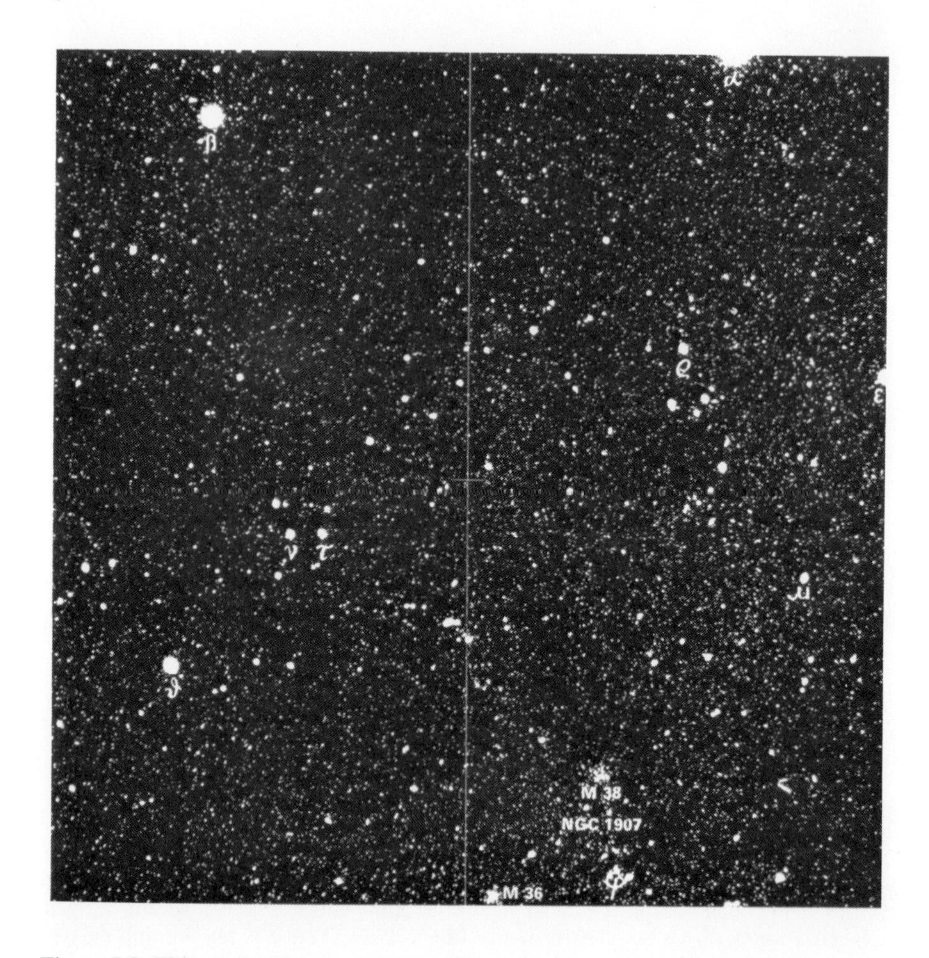

Figure I.3. This section from a star atlas shows an area in the Auriga constellation and represents part of the Milky Way. The three open star clusters M 36, M 48 and NGC 1907 can be seen in the lower part of the picture. Some of the brightest stars in the constellation are shown. Central coordinates are as follows: polar axis $\alpha = 5$ h 35 m, declination axis $\delta = +40°$. (Photograph: H. Vehrenberg)

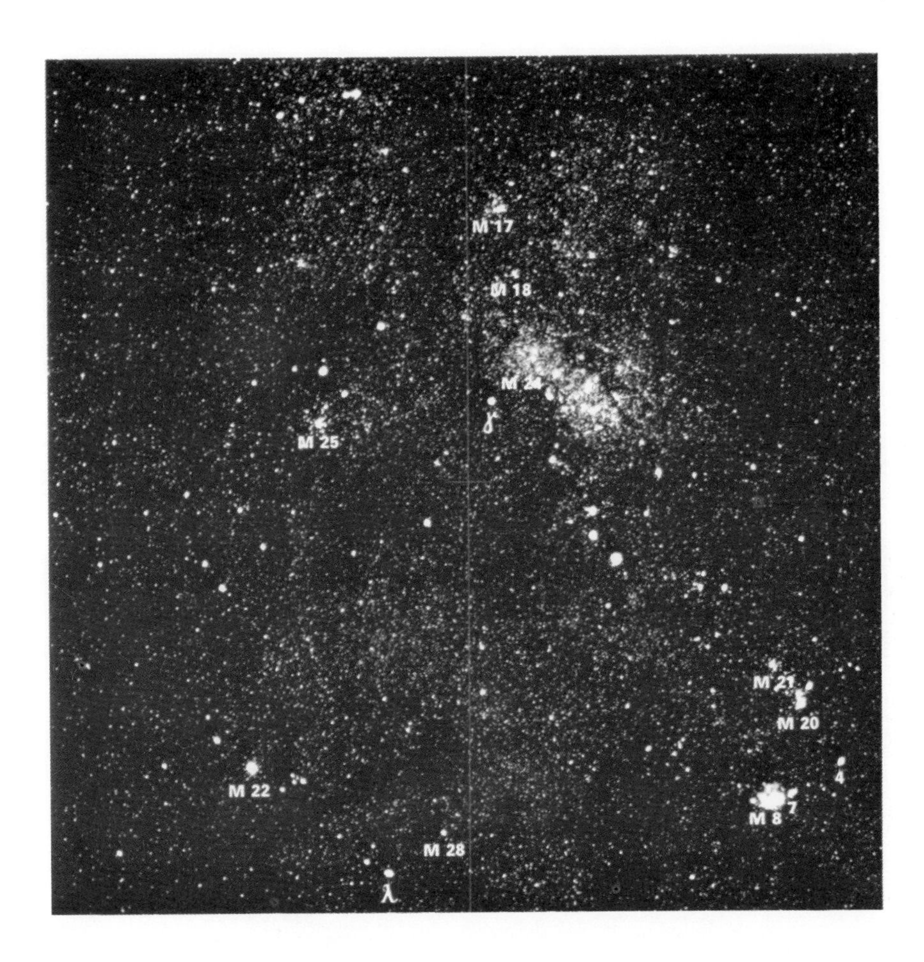

Figure I.4a. The centre of the Milky Way system lies in the Sagittarius constellation, so the wealth of stars found in this area is hardly surprising. There are also numerous open star clusters to be found here (this picture shows M 18, M 21, M 22, M 20, M 8, M 24, M 25 and M 28). M 27 is the Omega or Horseshoe Nebula—an H II area about 3 kpc away. The central coordinates are $\alpha = 18$ h 20 m and $\delta = -20°$. (Photograph: H. Vehrenberg)

Figure I.4b. The most significant feature of this view of part of the Sagittarius constellation is the 'star vacuum' or dark areas in the bottom right-hand section. Clouds of interstellar dust absorb the light from the stars behind them, thus giving the impression of a 'star vacuum'. Again, several open star clusters can be seen on the left. Central coordinates are α = 17 h 40 m and δ = $-20°$. (Photograph: H. Vehrenberg)

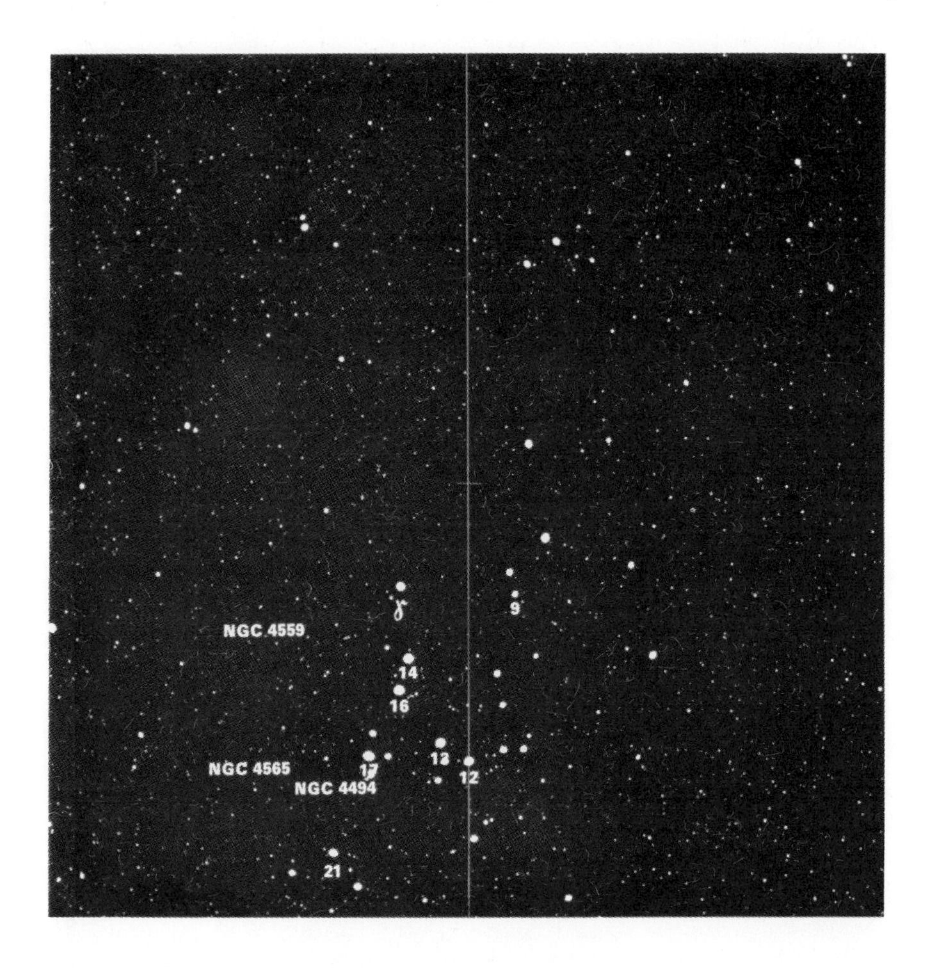

Figure I.5. The galactic north pole lies in the Coma Berenices constellation. At that point, we are looking outwards from the star system, perpendicular to the Milky Way. There are only a few stars in comparison with the previous set of pictures, but three extragalactic star systems are visible (NGC 4565, NGC 4494 and NGC 4559). This constellation also contains a galactic cluster with more than 1000 members. Most of the bright stars in the bottom part of the picture belong to an open star cluster which is only 80 pc away from us. Central coordinates are α = 12 h 20 m and δ = +30°. (Photograph: H. Vehrenberg)

Figure I.6. The apparent star density is particularly high when the line of vision runs along the galactic plane, for instance towards the Sagitarius constellation as shown here. The stars in this section are relatively uniformly distributed and three very distant globular clusters which are difficult to distinguish from bright stars can also be seen. M 69 is 6 kpc away and M 70 as far away as 20 kpc. M is short for Messier and M 69 means that it is the 69th non-stellar heavenly object listed in the catalogue published by the French astronomer Charles Messier in 1781. Objects referred to as NGC are from the New General Catalogue of Nebulae and Clusters, published in 1888. (Photograph: H. Vehrenberg)

II

A glimpse into the past

Although the unchanging pattern of fixed stars in the rotating celestial sphere did not fascinate the ancient astronomers quite as much as the moving planetary system did, nevertheless they did try to clarify certain aspects of it. The oldest concepts are tied up with gods and religious mythology, but they do represent the first attempts to establish some sort of order out of the apparently random star arrangements and to offer plausible explanations.

About 5000 years ago, the belief in Egypt and the Near East was that the sky was a large piece of fabric, tent or coat spread out over the Earth and decorated with stars. Alternatively, the sky was seen as a large river and the stars as boats moving about on it. Later, the sky was thought to stretch over the Earth—in the form of the Goddess Nut according to Egyptian cosmology.

This last concept was also held by learned Greek scholars. From Greece came the system synonymous with the name Ptolemy in which the Earth was the centre of the world with Sun, Moon and planets revolving around it and rounded off by the spherical layer of fixed stars. The stars were all the same distance away from the Earth, attached to a spherical surface which was viewed from the inside. This belief was upheld for centuries until Giardano Bruno's theory that the world was infinite and our Sun actually part of the star system surrounded by the same planetary formation. However, as the monk Bruno could not prove his concept, it remained pure speculation. In fact, he paid for it with his life; his philosophy was contrary to established Church beliefs of the time and he was sentenced by the Inquisition in Rome in 1600 and publicly burned. Although the Copernican world system was at the height of its popularity at the time, the fixed star concept remained intact for a while. Both Copernicus and Kepler thought of it as enclosing the world but Kepler's theory was that the fixed stars were contained in a small spherical layer and were all more or less equidistant from the Sun. He estimated this distance by assuming that Saturn's orbit was greater than the Sun by the same amount as the difference between the diameter of the fixed stars and Saturn's orbit. This gave Kepler a figure of 10,000 times the radius of the Earth's orbit from Sun to fixed stars, certainly a great distance but 27 times less than that of the nearest star by

Figure II.1. The ancient Egyptians thought that the Earth was encompassed by the Goddess Nut, who also bore the stars and was crossed over each day by the Sun barge. The Moon ship also followed a similar course. The picture shows both Sun and Moon barges (right and left, respectively). The Earth with its oceans is symbolized by the recumbent figure. (From O. Lodge, *Pioneers of Science*, Dover, New York, 1960)

modern calculations. However, in the seventeenth century, measuring techniques were not sophisticated enough to allow calculation of nearest star distances.

The great Dutch physicist and astronomer Christian Huygens, who also introduced the clock pendulum, thought up another method for estimating distance from the stars. He based it on the equality between Sun and stars, compared the brightness of the Sun with that of the stars and applied the well known law according to which the light source appears weaker the further the distance. Despite using correct methods, Huygens put the star Sirius at much too low a figure, *viz.*, 38,000 times the radius of the Earth's orbit, as it was and still is difficult to compare such a powerful light as that of the Sun with such a comparatively weak one as that emitted by Sirius. However, the fixed star theory was now somewhat in doubt. The next attack on it came in 1718, when Edmund Halley ascertained and pronounced that some stars, Arcturus, Betelgeuse, Sirius and Aldebaran had changed position in the heavens since the time of Ptolemy. Thus, the fixed stars were mobile. As far back as 1596, a keen astronomer from East Feiesia, David Fabricius, had observed a change in brightness in a star in the Cetus constellation, which was not at all compatible with the theory of absolute star constancy. The relevations introduced an element of chaos, because if the stars were not attached to a fixed point in the sky, they must be scattered about the universe at random and at that time there was no concept at all of a complete structured star system. Some ivory-towered

Figure II.2. According to Ptolemy's universe, the Earth was in the centre and the planets, Sun and Moon revolved around it. The Universe was enclosed by the firmament. This drawing is from the 1539 *Cosmographies* compiled by Peter Apian

scholars considered the matter and discussed many of the aspects in public. In England, Thomas Wright proposed his so-called 'Grindstone Theory' of the Universe in 1750. According to this, the Milky Way was a layered or disc-like structure of stars projected on to the celestial plane; our Sun was not in the central position but was nevertheless in the central plane. Looking from the inside of the star system, i.e. from our Sun outwards towards the 'grindstone' plane, many stars were visible, but that was just one aspect of the Milky Way. Looking vertically, only a few stars in the grindstone were visible, so these could then be viewed separately and clearly in the same way as those outside the Milky Way.

Just a few years later, in 1755, the young scholar Immanuel Kant published a short work entitled 'General Natural History and Celestial Theory or Research into the Constitution and Mechanical Origins of the Whole World Structure, Based on Newton's Law' (*Allgemeine Naturgeschichte und Theorie des Himmels oder Versuch von der Verfassung und dem mechanischen Ursprunge des ganzen Weltgebäudes nach Newtonischen Grundsätzen abgehandelt*) in Königsberg, East Prussia. This booklet contained details of the first really ingenious attempt to work out sun and star formation scientifically. The young philosopher based his work on the concept of gravity as described by Isaac Newton which was thought to apply throughout the Universe and into infinity. Kant argued that if this force affected the stars as distant suns, then after a certain time they would all collide; so, they must all move in large circles around a central sun. At this point, similarities to the planetary system become obvious. Just as the planets moved around the Sun in more or less one plane, then probably the many suns (fixed stars) would also move in one plane around the central sun and, according to Kant, this plane was in fact the Milky Way. How was it, then, that this movement, which must be obvious from the relative movement of other stars, could not be seen? Kant had an answer: if the distance between the fixed stars and the Sun were sufficiently great, then the apparent movement of stars as compared with the Sun would be slight (just as a fast, high-flying aeroplane seems to be travelling slowly when viewed from the ground) and, according to Kepler's Law, the rotation rate would be low if the stars were far removed from their central sun. At the time, the movement of some stars had been recorded over several years and Kant estimated that a star would take 1.5 million years to rotate around its central sun (the correct figure is 250 million years). Kant was a cautious man and made no definitive statement about the central sun itself. He wrote, 'perhaps it is up to future generations at least to find the area in which this central point of the fixed star system, to which our Sun belongs, lies, or maybe even to determine the precise location of this central body of the Universe, to which all parts are uniformly attracted.' He made a footnote here to the effect that his personal option was for Sirius, the dog-star, as the focal point. He also had another theory, which later turned out to be correct. For a long time, astronomers had observed bright nebula spots through the telescope, but no acceptable explanation for their presence had been forthcoming. Kant suggested that they represented clusters of stars which were similar to our Milky Way but which were so far away from us that they could not be seen in detail, even using the telescope. He saw them as distant 'cosmic islands'—star systems like our own but seen from the outside, as opposed to our own Milky Way, which we can see only from the inside. Kant subsequently changed his whole area of philosophical research and the small booklet was almost forgotten.

Six years after Kant's book, the 'Cosmological Letters' of Johann Heinrich Lambert, a former accountant from Mülheim in Alsace, were published in Augsburg. He proposed a similar concept of the star system to Kant, quite independently, but did not go into such great detail. However, he did go one step further. He saw the star world as a sequence of systems, the smallest of which formed the planet family, followed by the suns in the Milky Way ring, the so-called 'star ecliptic', and then on to other complete Milky Way systems which revolved around a gigantic centre, and so on *ad infinitum*. 'Thus, everything rotates around something else, the Earth around the Sun, the Sun around the central point of its system, this system itself around the central point it has in common with the other systems; and where does it all end?' The fact that these other systems were not visible was explained simply by the fact that not all systems and suns were illuminated. This was obviously not very satisfactory and the thought of systems of systems of smaller systems, all revolving around each other, is dizzy-making. However, these were all hypotheses, speculations and personal views which still had to be tested.

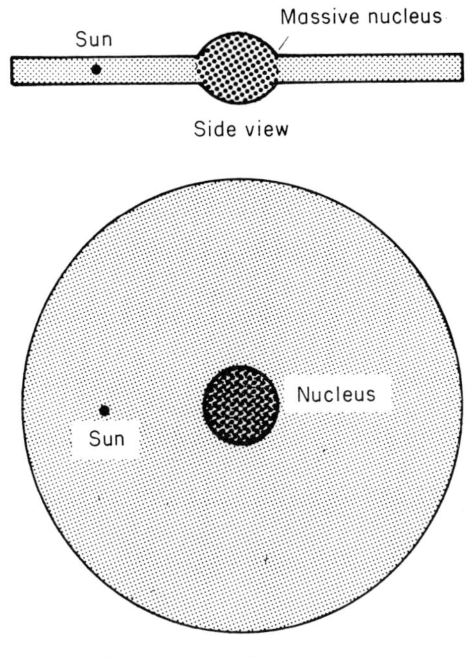

Figure II.3. According to the 'grindstone theory' devised by the Englishman Thomas Wright in 1750, the star system formed a disc shape and the Sun was to be found within the disc, but rather off-centre

Figure II.4. Side view of the Milky Way model by Friedrich Wilhelm Herschel. The deep bay on the right-hand side represents the visible split in the Milky Way in the Cygnus constellation. The drawing was made in 1785

At the same time, there was a man living in Bath in England who intended to make an accurate study of the structure of the sky by looking at the heavens more thoroughly. The man was Friedrich Wilhelm Herschel, known as the Prince of Astronomy for many years, and who made great strides forward in his pursuit of knowledge. He was born in 1738 in Hannover, the son of a military bandsman. He started off in his father's footsteps, but after the misfortunes suffered by Hannover during the first years of the Seven Years War, he left his regiment and moved to England where he earned his living as a musician. At the same time he began to experiment with building telescopes and to delve into astronomy. His knack for constructing new and better telescopes soon became apparent and before long he had made the most powerful mirror telescope of the time. By 1802, he had discovered 2500 nebula and more than 800 double stars and counted all the 4400 stars visible in the heavens with it. He not only discovered all these binaries, but also established that each star in the pair changed its position relative to the other over a period of time. He calculated that both stars in the Castor system in the Gemini constellation would move round each other over a period of 342 years. This showed that the law of mass gravity applied up there too and that the fixed stars were genuine suns. So much information based on one hypothesis! As an added bonus, the insatiable stargazer discovered the sixth largest planet in our solar system, Uranus, which rotates outside Saturn's orbit. But one of his most important research projects was to count the number of stars.

His diligence was well rewarded. First of all, Herschel established that the various celestial fields did not contain an equal number of stars. This was not a new concept, but star density calculations, based on the assumption that spatial density is similar throughout the star system, led to conclusions on the extension of the star system in various directions and hence the approximate shape of the fixed star system. This method can be illustrated by means of a terrestrial comparison. Let us assume we are in a small, lightly wooded area like a strip of woodland. As the trees in the centre of the wood are generally of similar density, we can see very many trees when we look along the longitudinal axis. However, the view along the diagonal is to open fields and very few trees can be seen. Tree density in the field of vision is, therefore, a measure of the extension of the wood. Herschel worked on the same principle. He counted the stars and took star density in the telescopic field of vision as a measure of the extension of the star system. However, the distance from the fixed stars was not yet known, so his figures were only relative. He took Sirius to be the 'standard distance' from the Sun, determined star density in the Sirius field and, using simple geometric calculations, specified the extension in this direction in Sirius widths. The same system was then used for other directions. This star count showed that our star system looks like a large lense from the outside, in which the Sun is in the centre, the maximum diameter is 1000 Sirius widths and the maximum depth a mere 100 Sirius widths.

Herschel's model shows a deep indentation representing the split in the Milky Way between the Cygnus and Aquila constellations due to an accumulation of light-absorbent dust grains.

The whole concept shows remarkable similarities with the accepted local system of today, and is all the more astonishing as Herschel knew nothing at all about the light-absorbent dust grains in interstellar space. The local system extends to 700–1000 parsec (pc). As Sirius is 2.67 pc from the Sun, Herschel's system is 2670 pc in size. Interstellar absorption factors would reduce Herschel's system somewhat. The unit of distance, the parsec (pc), used here is the one normally used in astronomy; 1 pc equals 3.26 light years or $30{,}857 \times 10^9$ km. The correspondingly larger unit is the kiloparsec; 1 kpc = 1000 pc.

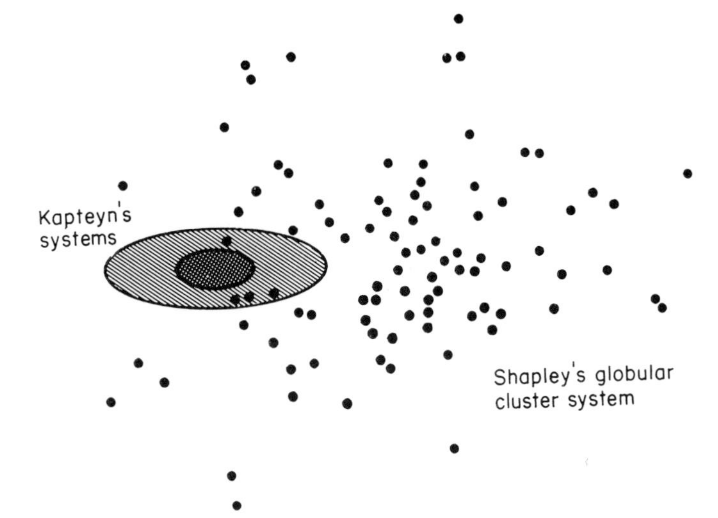

Figure II.5. Kapteyn's universe and Shapley's globular cluster system do not have a common centre. Shapley based his conclusion that the Sun was not in the centre of the star system, as Kapteyn had thought, on dynamic considerations, and also maintained that the system itself must be far larger than Kapteyn imagined

Nowadays, much of this speculation about our star system seems either elementary or untrue, or has little to do with the laws of nature. However, at that time, it represented great progress, as on the one hand speculation as to the structure of the heavens had been replaced by facts based on observation, and on the other it introduced statistical analysis into astronomical study. The long-lasting influence of Herschel's work is shown by the fact that 'Herschel's Heaven' was recognized in astronomy throughout the nineteenth century. This star system model was further developed and refined as a result of improved statistical analysis methods devised at the beginning of the present century, mainly by the Dutch astronomer J. C. Kapteyn. His star system was also lens shaped, with the Sun in the middle. Star density was greatest in the centre and became smaller towards the outside edges. The diameter of the system was 18 kpc and the greatest depth 3.6 kpc. At almost the same time as Kapteyn was engaged in his research, the American astronomer H. Shapley was tackling the problems of the star system structure from a completely different angle. He investigated the spatial distribution of smaller sub-systems, globular star clusters. He found that these star clusters were not concentrated along the Milky Way plane but formed a type of halo, the centre of which did not conform to Kapteyn's theories. There was no way of reconciling the two, so Shapley abandoned the Kapteyn system and proposed in its place a disc-shaped star system, 60 kpc in diameter, which is surrounded by a halo of globular star clusters and in which the Sun is within the disc, 15 kpc from the centre. This is very close to the modern concept.

Our ancestors knew that there were more than just stars in the sky. Even

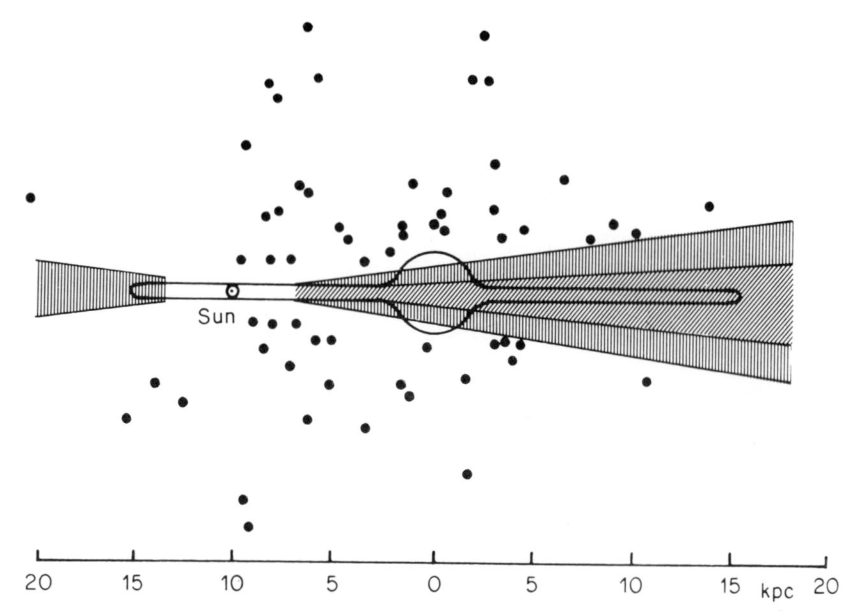

Figure II.6. Schematic side view of the star system as envisaged in recent times. A flat disc with a high concentration in the centre is surrounded by a halo of globular clusters. Only the non-shaded area can be seen clearly from the Sun. The light from the stars in the vertically shaded section is severely cut down by intestellar dust and the region which is obliquely shaded is virtually invisible. The Sun is in the moddle of the disc, but somewhat off-centre

before the telescope was invented, the bright nebula spots in the Andromeda constellation could be seen. Christian Huygens explained this as 'An opening in the sky, allowing a glimpse into illuminated regions beyond'. The Mayor of Danzig, Johannes Hevel, and the English astronomer Edmund Halley discovered several other 'nebula spots', but the Frenchman Charles Messier—named the 'comet hunter' by King Louis XV—was the first to compile a list of them; it contained 104 entries. This was considerably extended by Herschel, who brought out a catalogue containing 1000 entries of nebula spots and star clusters in 1786. He published one of a similar size in 1789 and a final one in 1802 containing a further 500 entries. The number of astronomical sightings increased rapidly as instruments improved. An 1890 catalogue lists 7840 and the modern Vorontsov-Velyaminov publication has a total of 39,000 bright star systems or spots.

Thomas Wright, Immanuel Kant and Johann Heinrich Lambert regarded these spots as cosmic islands, as complete star systems, but were not able to prove their theories. Even Herschel shared this opinion at first, but later rejected it as their increasingly diverse character became evident. He classified his sightings into categories ranging from pure gas to increasingly denser star clusters and concluded that they were certainly members of our star system. In spite of this, the cosmic island theory continued to fascinate astronomers and still had many supporters, in spite of no new evidence. This reached a peak in the middle of the nineteenth century when Lord Rosse of Parsonstown had a giant telescope with a 1.80 m diameter mirror built, and with this he discovered a new type of nebula, the spiral type. The theory was that this light scatter was made up of partially destroyed pieces of larger, previously rotating spherical star clusters. Once again, the concept of the existence of extra-galactic star or nebula systems outside our Milky Way fell into discredit when the gaseous nature of some nebulae was proved and their distance ascertained.

At the beginning of the present century, newer and larger telescopes were erected at specially favourable sites and the true solution was pieced together. First of all, two basic types of nebula were isolated; one which belongs to our star system and an extra-galactic one which lies far beyond our Milky Way. In 1926, using the 2.5 m mirror telescope at the Mount Wilson observatory in California, the American Edwin P. Hubble finally obtained proof that lens and spiral-shaped extra-galactic neubla were distant star galaxies like our Milky Way. He was in fact able to isolate many individual stars along the edges of the Andromeda Nebula. Later, the centre of this great star system was seen as a series of dense star clusters through the 5 m telescope on Mount Palomar. So, there is no longer any cause to doubt that the nebula spots of extra-galactic star systems are related to our Milky Way. They are certainly many and varied: narrow and broad spirals, spherical, elliptical and some highly irregular and odd. If our Milky Way system is a comparable one, what type does it belong to? The answers provided by Herschel and Kapteyn to this question no longer suffice. The structure of our star system is the object of current galactic research—an important branch of astronomy and astrophysics, and it is this area which we intend to cover in the following chapters.

III

Looking over the astronomer's shoulder

Light as Information Media

Man is not just a questioning and inquisitive being, he also feels a strong compulsion to communicate any experience or information acquired. He is a social animal in the best sense of the word, assuming, of course, that he does in fact have the means to report and communicate. Man has the ability to exchange information. Even when separated, two people are still able to carry on a dialogue. The information is committed to paper, put in the post and the recipient can eventually read what has been written. Using the modern jargon, we would say that sender A has encoded (written out) the information, fed it into the input/output media (paper) and sent it to recipient B, who has then decoded (read) the message. Scientists are very keen to hold a dialogue with nature. For example, a chemist or a physicist will attempt to provoke nature into providing answers to his questions in the laboratory. Even if these answers are encoded, he can still 'hear' (i.e. recognize) them in the laboratory. Astronomers are up against a totally different set of circumstances. Their area of research is far removed, so that objects can be neither 'provoked' nor 'heard'. Therefore, they must learn how to interpret correctly the relevant messages conveyed by the 'star post'. The latter is simply the light emitted by the stars and the message is actually contained in the light, i.e. is represented by the special qualities of the particular star light under examination. We could say that all the information in stars and star systems is contained in the emitted light, which acts as an information medium. Let us now 'look' more closely at that light.

At school a child learns that light is an electromagnetic wave. This is where the learning process ends unless a concept of electromagnetic waves is understood. A plate condenser can be used as a simple example. When a voltage is applied between the two plates, an electric field is set up, i.e. the space is changed such that a force is exerted on a charged particle. If an alternating current voltage is applied, then this electric field is created and destroyed in the same cycle. If a resonant circuit is now substituted, thus giving a coil as well, a magnetic field is created and destroyed on the coil in addition to the electric field on the condenser. An electromagnetic exchange field has been set up. By pulling the condenser plates apart, the size of the electromagnetic field can be increased until it finally covers the whole area. Now the unit acts as an aerial, relaying the electromagnetic waves. When the voltage variation is relatively slow—of the order of about a million times per second—then a radio wave is radiated. It is of relatively low frequency or, put another way, has a long wavelength.

Visible light consists of electromagnetic waves of very short wavelength or very high frequency (the field oscillates 5×10^{11} times per second). Such a transmitter cannot be constructed by electronic means alone, but atoms and molecules are themselves transmitters and waves radiated from atomic nuclei are even shorter. The whole range of electromagnetic waves is known as the electromagnetic spectrum. The special qualities of each star are imprinted in this spectrum. Thus, some broader wavelength ranges can be radiated either very strongly or very weakly. If a narrow wavelength range disappears completely or is very weak, then we speak of an absorption line; on the other hand, a particularly strong signal gives an emission line in the spectrum. In addition, light or, more generally, electromagnetic waves, have a further special quality. The electric or magnetic field is not just characterized by its force but also by the direction in which a test charge would move. With normal light, this direction changes at random. However, when the direction remains constant and only the sign changes, thus reducing or increasing field strength, we say that the light is polarized.

Astronomers study light in various ways in order to be able to decode star information:

— the direction from which the light emerges defines the location and movement of the star vertical to the optical path (proper motion);
— the brightness in all spectral areas gives information on the total energy emitted by the star;
— the chemical composition of the stars, temperature and density, to name but a few factors, are all revealed by the star's spectrum; the spectrum is a star's actual code key;
— polarization measurements are used to determine the magnetic field of the stars and it can also reveal what has happened to the light on its journey through space.

Astronomical Range

No-one would find it difficult to plot out a distance of 10 or 20 m. It is slightly more difficult to assess accurately the distance between two towns, and an outsider would find the task of assessing the distance between two continents, e.g. the American and European coasts, to within several metres, very daunting indeed. The problem of measuring distances in space is no less

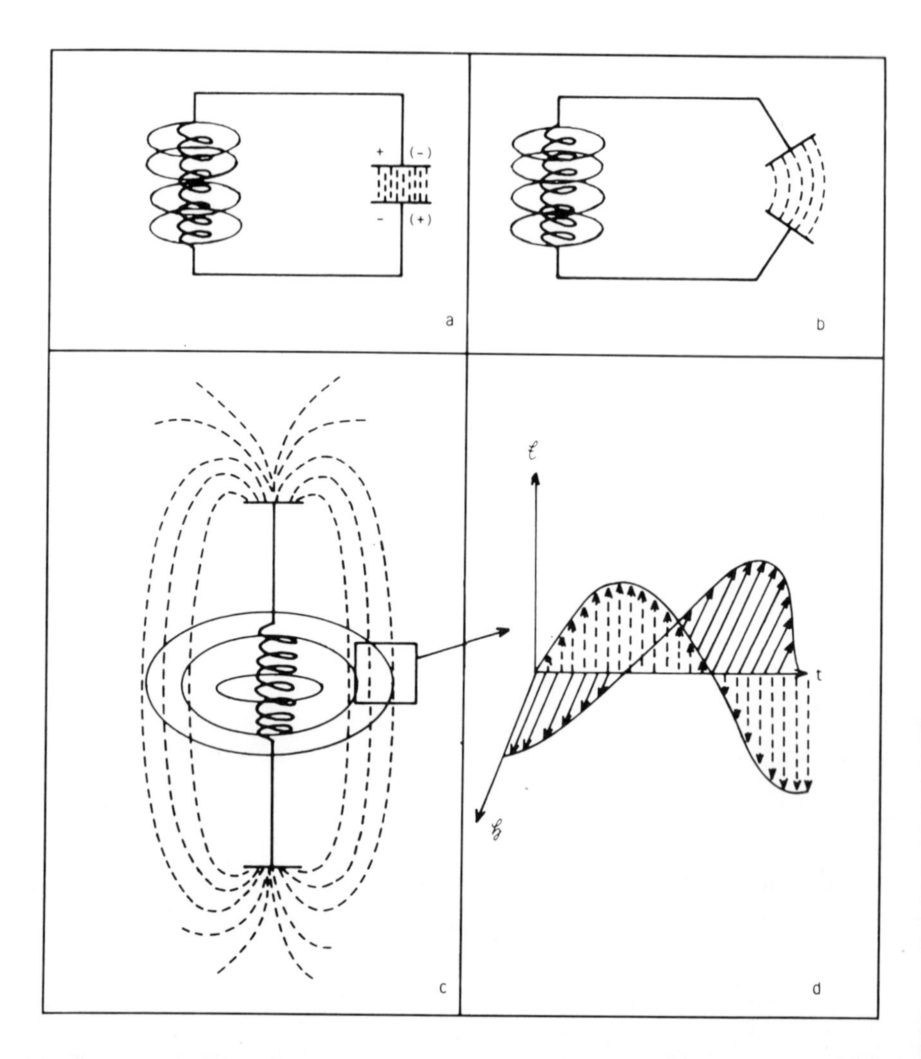

Figure III.1. A simple resonant circuit made up of a plate condenser and a coil: when an alternating-current voltage is applied between the two condenser plates, a rising and falling electrical field is created and an oscillating magnetic field is formed around the coil. If the condenser plates are pulled apart, the electric field extends over a much larger area until finally the resonant circuit is at full capacity (dipole antenna) and the whole area is under the influence of both electric and magnetic fields. The two oscillating fields overlap each other and send out an electromagnetic wave from the resonant circuit. When sufficiently far away from the resonant circuit (antenna), the curvature of the field lines is insignificant and the wave front forms a plane. The electric and magnetic fields are vertical to each other and are phase-lagged, i.e. the magnetic field strength reaches its maximum as the electric field decreases to zero. If the field strength is represented by arrows of different lengths, then a sine curve emerges. The electromagnetic wave represented here is polarized as both electric and magnetic field strengths oscillate in one plane

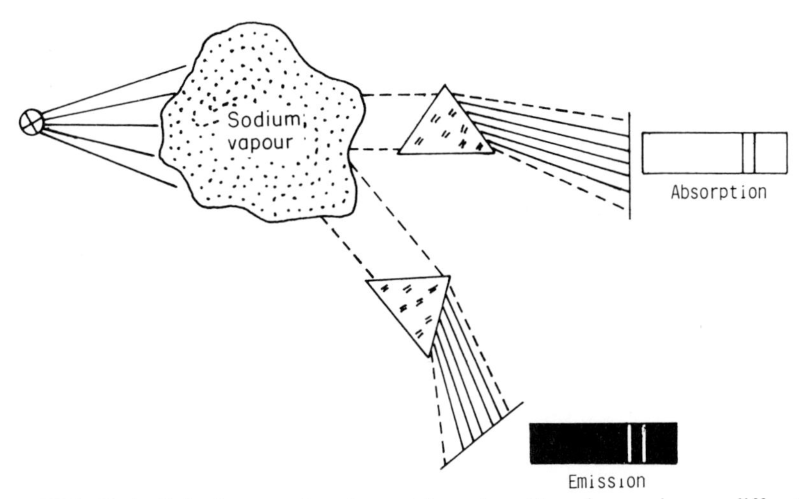

Figure III.2. If the light from an incadescent lamp is split up by a prism or diffraction screen, a continuous spectrum is obtained, i.e. no wavelength interval stands out as being of especially high or low intensity. When a suitable gas, e.g. sodium vapour, is found between the lamp and the prism, this absorbs some wavelengths and these then fail to appear in the continuous spectrum; we observe one or several absorption lines. However, if the sodium cloud is viewed on its own, then there is no continuous spectrum and only the bright sodium lines appear (emission lines)

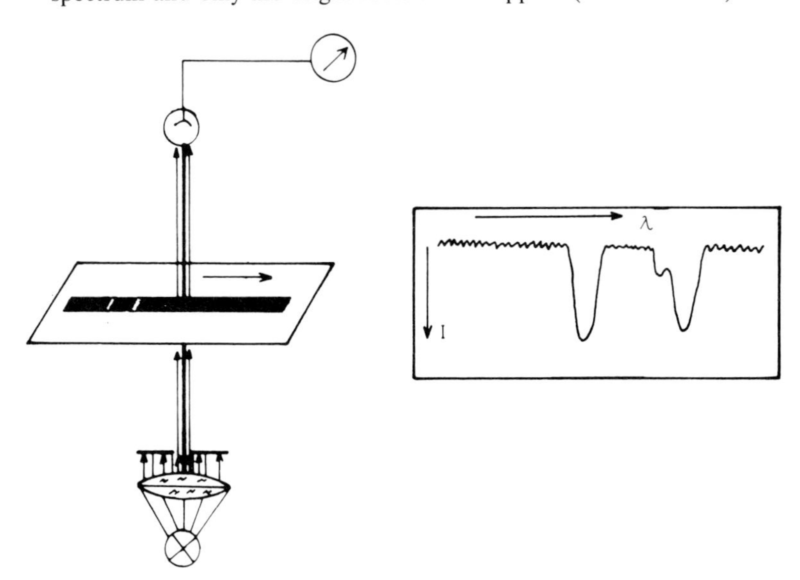

Figure III.3. The photographed spectrum can be evaluated more satisfactorily by recording it. To this end, a sharply focused beam of light is passed through the photographic spectrum and the intensity of the beam is measured with a photocell. This transmitted intensity is dependent on the density of the photographic plate. If the photographic plate is then passed slowly through the beam of light, the density (which is a measure of intensity) is shown as a function of wavelength and the spectral lines emerge very clearly

difficult but has the added disadvantage that much of it is inaccessible to Man. However, methods for determining the distances between far-off objects have been devised and this is a highly important facet of astronomy as distance plays a significant part in assessing star properties.

Astronomers have based one of their most effective methods on the principles of geodata or land measurement. The object in question is looked at from two different points and the corresponding angle to a zero line is measured. As long as the distance between both measuring points is known, then the distance from the object can be calculated from the angles measured, without difficulty. This is called the trigonometric method as it is based on triangular measurements. However, problems do occur when the process is adopted for astronomy. Even the greatest distance possible between two points on Earth is too small for most distances in space, and angle measurements could scarcely be defined. However, the astronomers came up with an answer: the angle to the star was to be measured on a certain day, then again 6 months later. In this half year, the Earth would be regarded as a spaceship and cover double the distance between the Sun and Earth, seen as an 'air-line', from the spot at which the original measurements were carried out. This distance is about 12,000 times greater than the largest possible figure on Earth and a star which is 3.26 light years away from us can be seen between the angle of 2 arc seconds (one arc second is a 3600th part of one degree; a 5 pence piece would be seen between this angle from 5 km away). The distance of 3.26 light years is known as a parsec and is a normal astronomical term of measurement. Using the whole Sun system as a spaceship provides an even broader basis. We now know that the Sun moves at a rate of about 30 km/s within the neighbouring star reference system, in the direction of the constellation Hercules. By measuring the alignment with the star at a certain point in time and then again several years later, both distance and angle between measuring points are then known and can be used to calculate how far away the star is. The only snag is that the star may have moved in the meantime and this movement is very difficult to separate from that simulated by the Sun's motion, so the method is used for statistical purposes, for preference.

In spite of the many refinements in measuring techniques which have been developed in the intervening years, trigonometric methods are still suitable only for the immediate area around the Sun. Astronomers have therefore devised another method for ascertaining distance. It is a long established fact that two equally bright light sources which are different distances away from us do not seem to have an equal level of brightness. There is a fixed relationship between brightness and distance, so that once true brightness is defined, distance can then be calculated. Astronomy has devised the concept of absolute brightness to represent the true figure; if the star is 'placed' 10 pc from the Earth, then the apparent, measurable brightness is equivalent to the absolute value. The latter is, therefore, a measure of the star's luminosity. Absolute brightness can sometimes be calculated even if distance is unknown. Accurate research into the spectrum provides such a possibility. These

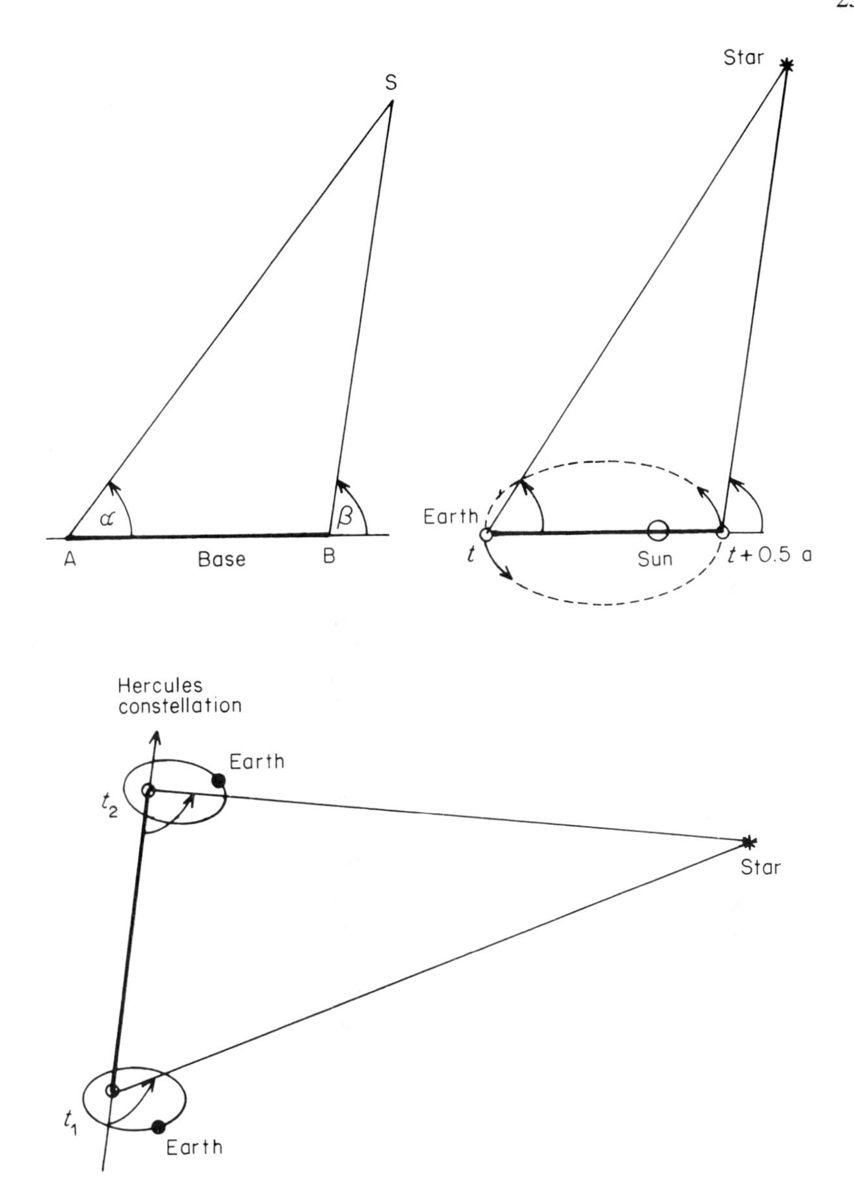

Figure III.4. Trigonometric distance calculations are carried out by measuring both base angles of a triangle with known base. As any Earthly distance is too small for astronomical calculations, the diameter of the Earth's orbit is taken as a base and the angles are measured at two points in time separated by a period of 6 months. An even larger base line is obtained by using the movement of the Sun for the same purpose. As the Sun moves towards the Hercules constellation at a rate of about 20 km/s, two measurements with a 5 year gap in between will give a base of more than 10 times the diameter of the Earth's orbit. However, as we do not know whether or not the star in question has been at rest this time span, this latter method is used mainly for statistical purposes

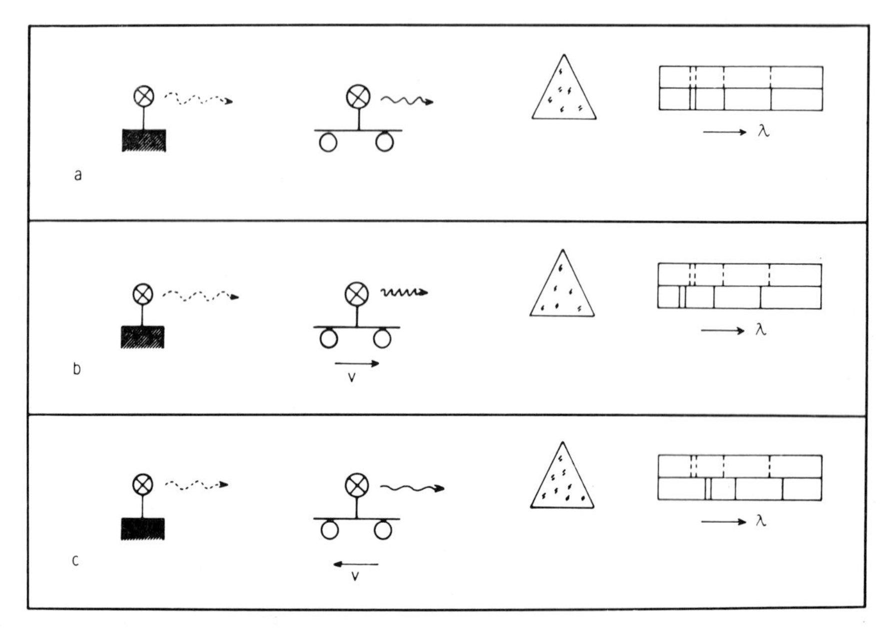

Figure III.5. If the observer looks at a relatively mobile light source, the Doppler effect causes spectral displacement. This displacement can be clearly seen and measured on a line spectrum and also helps to determine the velocity of the light source. If the light source (in this case a gas-filled lamp with a characteristic emission line spectrum) moves closer to the observer, then the lines move towards smaller wavelengths (b) (blue displacement) than those in the comparison spectrum of an immobile light source. Conversely, there is a shift towards longer wavelengths (c) (red displacement) as the light source moves away from the observer

photometric methods for determining distance obviously had to be standardized first of all on stars for which distance had been calculated by trigonometric measurement. The group of Hyades stars was and still is used for this purpose and astronomers are always having to re-measure its distance, using the utmost care each time. Over the last few decades, much of the fixed star world has been investigated by photometric methods.

An Important Discovery Made by a Teacher (Doppler Effect)

In 1842, a grammar school teacher and later Professor of Physics, Christian Doppler, published his tests on the frequendy changes of mobile wave transmitters. Either a source of light or a source of sound can act as a 'wave source'. In the case of an acoustic Doppler effect, in which a sound source moves, frequency changes are expressed by a change in pitch, whereas an optical Doppler effect is accompanied by spectral-line displacement. If the sound source and the observer remain still, then the latter hears a sound of specific frequency. However, should the sound source move towards the observer the sound becomes higher pitched, to a level dependent on the velocity of the sound source. The reverse is true when the sound source moves away from the observer, so that the pitch is lower than that of the sound emitted during the state of rest. the same principle applies for a moving light source. Assuming that the light source emits light of only one frequency or wavelength (i.e. an ideal, sharply defined emission line), then the spectral line will move towards the violet part of the spectrum (frequency increase) as the observer approaches, but towards the red part as he moves away. Christian Doppler applied this knowledge to astronomy; he rightly assumed that the great velocities which occur in space would give rise to the same effect (the degree of line displacement is a function of the ratio between the velocity of the light source and that of the light itself, and on Earth this figure is generally very low). He attributed the different colours evident in double star pairs to this same phenomenon. Although we now know that this explanation of binaries is incorrect, the Doppler effect is a very important concept in astronomy as measurement of spectral-line displacement in the star spectrum yields information on the radial velocity of the stars (i.e. velocity along the line of vision), which in turn is used to calculate many binary orbits, thus providing data on star rotation. A special advantage of the Doppler effect is that only *one* spectrum is required to determine radial velocity and the rate can be given direct in km/s, even when distance is unknown. Modern star spectrographs can determine the velocity of brighter sun-type stars to within 1 km/s and of weaker stars to within 5 km/s.

IV

Are there really only stars?

The 'Empty' Space between the Stars

When there are just a few drops of liquid left in a litre bottle, we normally claim that the bottle is empty. However, to be accurate, about one thousandth part of the bottle is still occupied by the liquid. Now we could carry this analogy over into the realm of space. Long-exposure pictures of celestial areas around the Milky Way show stars covering virtually the whole surface. However, distance plays a part and once this is known, then the surface density of stars can certainly be used as a basis for calculating spatial density; this in turn will then show how 'full' space actually is. The result is astonishing: if a cube measuring 2 pc along the edges were to be cut out of space there would be an average of one star in each space box. The 'empty' litre bottle would still be 10 billion times 1 billion times more full than the celestial cube. This is certainly way beyond comprehension and, in order to illustrate spatial ratios rather more simply, let us imagine that the world could be reduced to a scale of $1 : 10^9$, such that the diameter of the Sun was a mere 15 cm; then the 1 mm Earth would rotate around it 15 m away, the nearest star would be 5000 km from the Sun, the next one about 4000–5000 km, and so on. Obviously the most salient feature of space is its emptiness, and indeed this is not an unknown concept, especially in the field of atomic physics. A further comparison can be made between the fixed stars and this 'small world' by reducing the size of the heavens and increasing that of the atoms.

First Comparison: Fixed Stars–Gas

We know that gases are formed from molecules or atoms which move in a random way and are continuously colliding. If the size of these molecules is increased to 15 cm spheres, then under normal conditions they would move about 85 m along before two of them finally collided. Using the same scale of reference, the stars are about 4000 km from each other, so the 'star gas' surrounding the star world is extremely dilute and any experimental physicist

Figure IV.1. The Orion Nebula (M 42 = NGC 1976) is a large diffuse gas nebula in Orion's sword belt. The smaller gas nebula M 43 can also be seen. The gas in this Nebula is stimulated to emit light by hot, young stars. Stars are still being formed in the Orion region. Estimates put the age of the Nebula at about 300,000 years. The dark stripes at the bottom are due to dust masses which also form part of the Nebula. (Photograph: Lick Observatory)

would be delighted to have such a vacuum in his laboratory. Obviously, stellar encounters rarely occur under these circumstances and none has ever actually been observed.

Second Comparison: Fixed Stars–Atom

Now, let us compare the almost empty 'space bottle' with the 'atom bottle'. Physicists say that atoms consist of 'nothing' for the most part and in practice only the minute nucleus contributes anything to the mass. So, how much space is actually taken up by the mass if our bottle is filled with hydrogen gas, for example, under normal conditions? Brief calculation shows that only 1 part in 2.4×10^{-21} of the 'atom bottle' is actually filled—an inconceivable figure? Perhaps it can be better grasped by saying that a 100 km balloon filled with hydrogen gas could in effect be forced into the bottle. This 'atomic vacuum' is comparable to the fixed star vacuum even though the latter is still about 20 times more empty in the vicinity of the Sun. Although the stars are, therefore, very far away from each other and take up only a tiny amount of space in comparison with the total volume, the spaces between them are certainly not completely empty and lacking in interest. The concept of a vacuum between the stars has always been based on lack of matter, as obviously the light from the stars radiates all over. Apart from this, there is always a magnetic field and a gravity field and elementary particles such as electrons, protons and even whole atomic nuclei are continuously shooting about in space at great speeds. Considerable amounts of energy are present in the radiation and magnetic fields and in the cosmic radiation bands. This interstellar space, as we shall call it from now on, has become of increasing interest to astonomers over the past few decades.

Dark and Bright Clouds

The fact that there are both old and young stars, that not all of them live for the same length of time and that some are even now in the stages of formation has been known for several decades. Consequently, the raw materials for this production must be somewhere in space, and interstellar space seems an obvious place to search. However, a few more concise details should be given first.

With the naked eye, only a few nebula spots (e.g. those in the Andromeda, Orion or Traingulum constellations) or stars can be seen and the areas in between seem to be in complete darkness. The telescope gives a more complete picture. Bright, irregular, very odd looking nebula spots have been discovered which even the telescope cannot break down into individual stars and which are obviously far from being distant cosmic islands. When the light from one of these spots is examined spectroscopically, many emission lines can be seen. However, that is just proof that the nebula is composed of gas. By far the best example of this type of diffuse or emission nebula is the Orion Nebula.

Another type of bright nebula was found in the Pleiades constellation in Taurus. Several stars were actually embedded in the nebula and the light from it had the same properties as the stars. Obviously, the light from the stars falls only on small solid particles—dust grains—and is then reflected. This type of nebula is called a reflexion nebula. The dust grains are found in space in other forms. In comparison with their surroundings, many parts of the sky seem to have stars missing, there are real holes, or 'dark nebulae' to use the correct jargon. The layers of dust accumulate in front of the star and absorb, i.e. block the light, thus giving the impression that no star is there. This accumulation of dust grains is known as dark nebula. In placed where there is only a little dust in front of the star, the starlight appears red as the small dust grains absorb or diffuse the blue light more readily so that the light which actually penetrates through tends towards the red end of the spectrum. This selective absorption effect can be measured and used to determine dust quantities.

Astronomers found further circumstantial evidence for the existence of dark or not directly visible interstellar matter. Accurate observation of the heavens reveals that stars come in pairs as well as singly. Each one of the pair rotates

Figure IV.2. The Trifid Nebula is found in the Sagittarius constellation (M 20 = NGC 6514); it is a diffuse gas nebula 700 pc away. (Photograph: Lick Observatory)

around the other and the Doppler effect reveals this in the spectrum—the absorption lines wobble to and fro. Some double star spectra do produce straight lines, but these could not possibly have been from the stellar atmosphere, they must have been imprinted on to the star spectrum by an otherwise invisible accumulation of gas in front of the star. Since this discovery, the same type of interstellar absorption line has been found in many star spectra (and not only from double stars) and can thus be regarded as a rough representation of the 'chemistry' of interstellar gas.

Figure IV.3. Open star cluster Pleiades in the Taurus constellation. More that 250 stars belong to this cluster, formed from a large gas and dust cloud about 4.2 million years ago. It is about 140 pc away. Some of the stars in the cluster are embedded in interstellar matter with a high dust content. The dust particles, which are only 10^{-5} cm in size, reflect the starlight, thus producing the nebula glow. The filament type structure is typical of these type of reflexion nebulae. (Photograph: Lick Observatory)

Gas Nebulae as Interstellar Fluorescent Light

Interstellar space is, therefore, not only filled with field energy but also contains actual matter, as manifest by the light and dark clouds. This interstellar matter is the raw material for star production and its physical condition is responsible for the 'when, how and where' a star is born. The most conspicuous interstellar space phenomenon is, without question, the bright diffuse nebulae. Let us examine how such a nebula functions.

Observation of many gas nebulae showed that they are always linked with one or several very bright hot O or B spectral type stars and the gas in the nebula consists predominantly of hydrogen and helium plus small amounts of heavier elements. One important property of the O and B stars is the high UV content in their total light. This UV light is so rich in energy that it ionizes most elements. Hence the hydrogen in the gas nebula is almost fully ionized, i.e. the atoms are decomposed into electrons and nuclei (ionized hydrogen is known as H II, so such gas nebulae are also called H II regions). The relatively free electrons are trapped by the protons (hydrogen atom nuclei) at a higher energy level, thus giving out a light wave and, after a short period, the highly charged

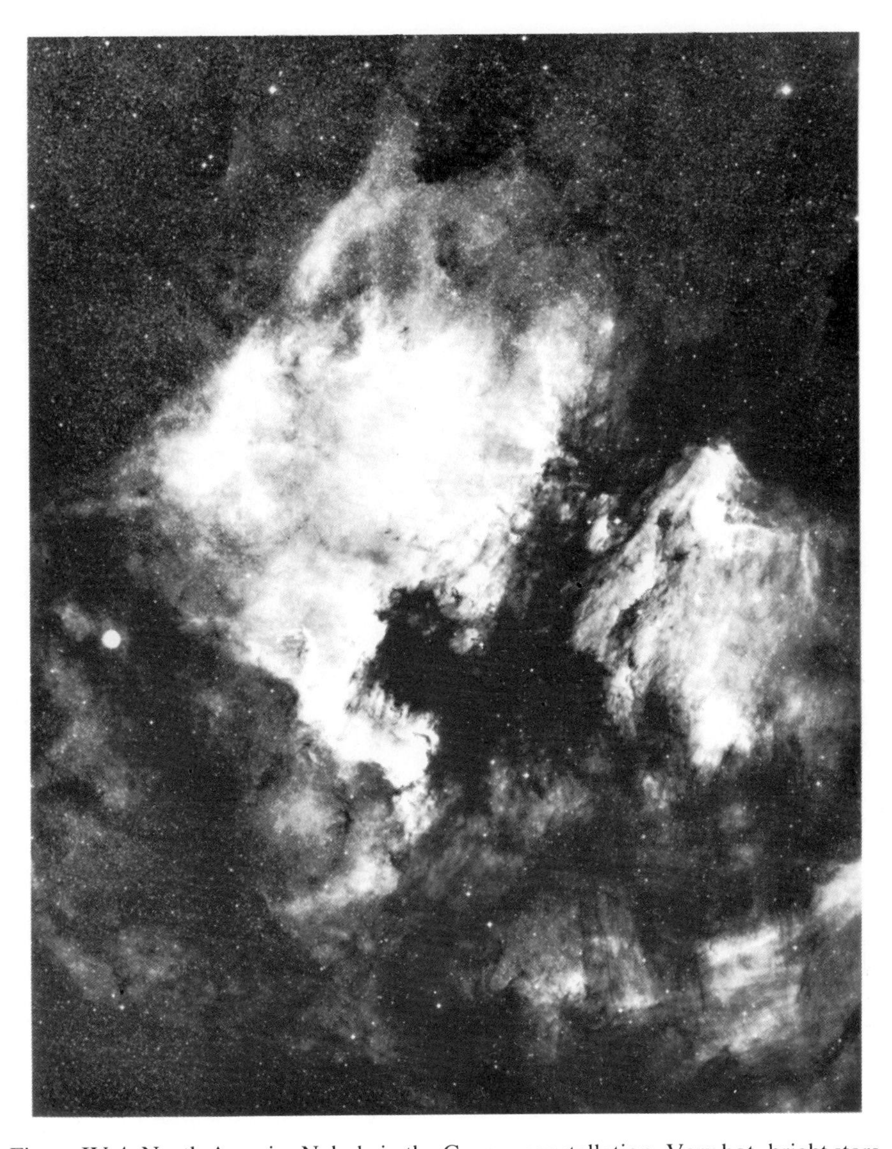

Figure IV.4. North America Nebula in the Cygnus constellation. Very hot, bright stars act as a light-stimulant to the hydrogen gas, which produces a particularly intense glow in red light. The characteristic shape of the Nebula is due to dark dust matter and is emphasized even more by the relative 'star vacuum' on the right-hand North American coast and in the Gulf of Mexico. If the red light in the hydrogen is filtered out and the region then photographed, the Nebula itself is no longer in evidence although the North American contours can still be detected as a result of the 'star hollow'. This provides proof of the existence of interstellar dust. (Photograph: Mt. Palomar Observatory)

Figure IV.5. The Horsehead Nebula in the Orion constellation to the south of Orion's Zeta star is a particularly impressive example of dark, light-absorbent matter in front of a luminous gas mass. The remarkable aspect of this Nebula is the glow around the dark edges. (Photograph: Mt. Palomar Observatory)

hydrogen atom is converted to a lower energy level, resulting in a further light wave emission. The hydrogen atom is ionized in a single step, but the conversion back to the basic atom takes place in stages. In this way, the high-energy UV light is converted to visible light and this is the principle of fluorescent tubes. The electric discharge produces UV light waves which are converted to light on reaching the fluorescent substance in the glass walls.

As well as the 'UV transformer', collision processes contribute to nebulae light. In these areas, the density is so high that frequent atom or ion collisions occur, in which the partner is stimulated to a higher energy level. The energy required to do this is drawn from the particle's kinetic energy and the subsequent transfer to a lower energy level again results in emission of light. Tests on all these processes within the nebulae have yielded information on density, temperature and composition. Hence we know nowadays that there are relatively small emission nebulae which contain only few solar masses of material in their total mass and gigantic H II regions which contain 15,000 or more solar masses (the greatest recorded figure is 100,000). Even the density of the nebulae fluctuates between a few and several thousand hydrogen atoms per cm^3. On the other hand, the temperature is almost always around 10,000 K. Its proximity to the comparatively very young O and B stars makes it highly likely

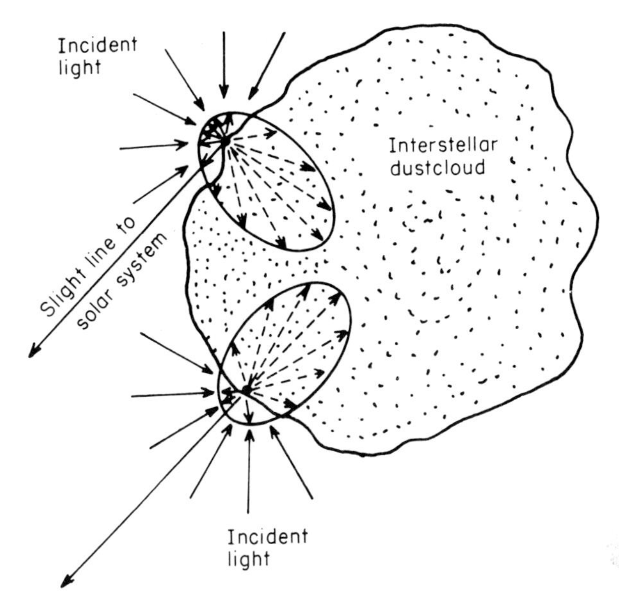

Figure IV.6. The dust particles around the edges of a dark nebula scatter more light in our direction than the particles which we can see direct. Therefore, the dark clouds appear as structures with bright edges. In the diagram, the incident starlight is represented by solid arrows and the dust-scattered light by broken arrows

that the nebula is made up of matter either left over from or still being used for new star formation.

Dust Clouds

Many will know how to make a light ray visible or have seen the light beam from the Sun's rays as they stream through a small window into a dark room. The beam of light is visible in this case because the dust particles floating about in the air are radiated only in that particular area, they reflect the light in the direction of the observer and thus radiate a brightness all of their own. Exactly the same process occurs in a reflexion nebula and in certain circumstances reflected light measurements can give information on particle size and quantity and even include yet another so-called material property, the reflection factor or albedo. However, interstellar absorption is just as important a source of information on interstellar dust.

Star light is scattered on the widely dispersed interstellar dust particles and dispersion is greater the shorter is the wavelength and the bluer is the light. Thus, on a long journey through a dust-filled space, much more short-wave than long-wave light will be dispersed, leaving mostly the long-wave red light left over (the red light of the morning and evening Sun is a good example of this phenomenon, except that the light itself does not fall on dust particles in this case). Particle size and shape are of decisive importance for this selective

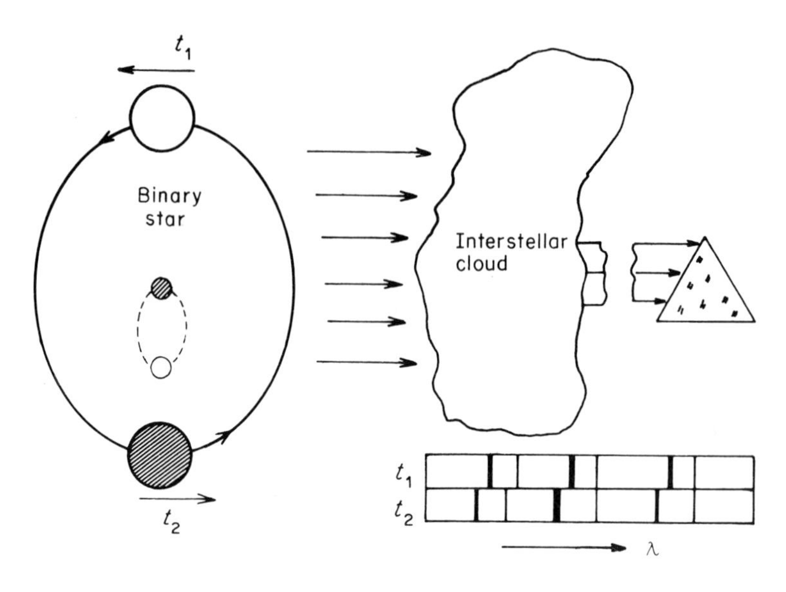

a

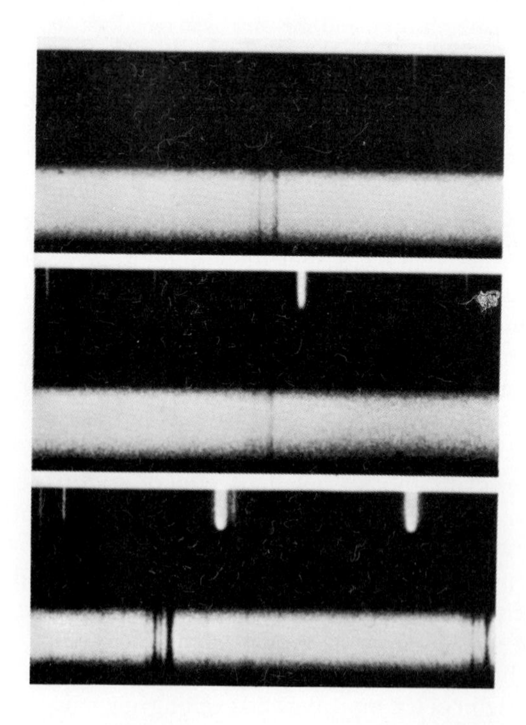

b

absorption process and the refractive index—yet another material property—also plays an important part.

A summary of all the visible and circumstantial evidence on interstellar matter gives the following picture: tests on interstellar absorption lines showed that the chemical composition of interstellar gas is essentially uniform throughout and is the same as that of many stars; the bright nebulae are not very different in this respect from the general interstellar clouds. However, temperature and density figures for the diffuse nebulae are much higher. A 'normal' cloud has a density of only 1–100 particles/cm^3 (an inconceivable vacuum by Earth's standards), whereas that of many light nebulae is up to 1000 times greater. In general, dust and gas are well mixed in interstellar space, although the dust content is obviously not uniform in all clouds. However, the dust level in interstellar matter is a mere 1%. In spite of a great deal of research, very little is known about the nature of these dust particles. The absorption and reflection of starlight actually observed is probably caused by just a few 10^{-5} cm large silicate particles (similar to the Earth's variety) and some graphite ones. As absorption is a function of wavelength, the particles cannot be much bigger than the wavelength of the light, as larger ones would simply hide the light and give rise to an absorption process totally independent of wavelength. Some dust accumulation form dark clouds or nebulae. Large dark nebulae contain up to 100 solar masses of dust whereas the small but dense dust accumulations, known as globules, contain only a few hundredths of a solar mass of dust. A cube with edges 30 m long, enclosing a dust particle inside a 'normal' dark cloud, would be correspondingly reduced to 5 m long sides inside a globule. In spite of these 'tidy' ratios, based on terrestrial dust density, both dark cloud and globules are virtually impervious to light and thus make it seem as though there are empty spaces or holes in the sky.

So far, we have spoken about light and dark clouds, interstellar absorption and interstellar lines as related to interstellar matter. However, there is a

---

Figure IV.7a. The non-luminous interstellar matter also makes its mark on the starlight passing through. 'Stationary lines' show this very clearly. In a suitable double star pair, one star moves away from us at time t_1, and as a result of the Doppler effect the spectral lines of the star spectrum are then displaced to longer wavelengths (red displacement). At a later point in time, t_2, it comes back towards us again and the lines are now displaced to shorter wavelengths. Thus, the spectral lines of a star 'rock' to and fro. On the other hand, the dark, narrow absorption lines produced in the interstellar medium (mainly those of sodium and calcium) remain stable—they are stationary

Figure IV.7b. Interstellar gas clouds between us and a star absorb part of the starlight and thus give the typical absorption lines in the star spectrum. In contrast to stellar lines, these interstellar lines are very sharply defined and often split into several components. Each of these components comes from a cloud and has a specific radial velocity. A particularly fine example of these interstellar lines is found in the spectrum of the ε Orionis star. Five components show the H and K lines of simple ionized calcium and neutral sodium (D-lines). They come from five different interstellar gas clouds with a radial velocity of +3.9, +11.3, +17.6, +24.8 and +27.6 km/s, respectively. At these velocities the clouds are moving away from our line of vision. (Photograph: Mt. Palomar Observatory)

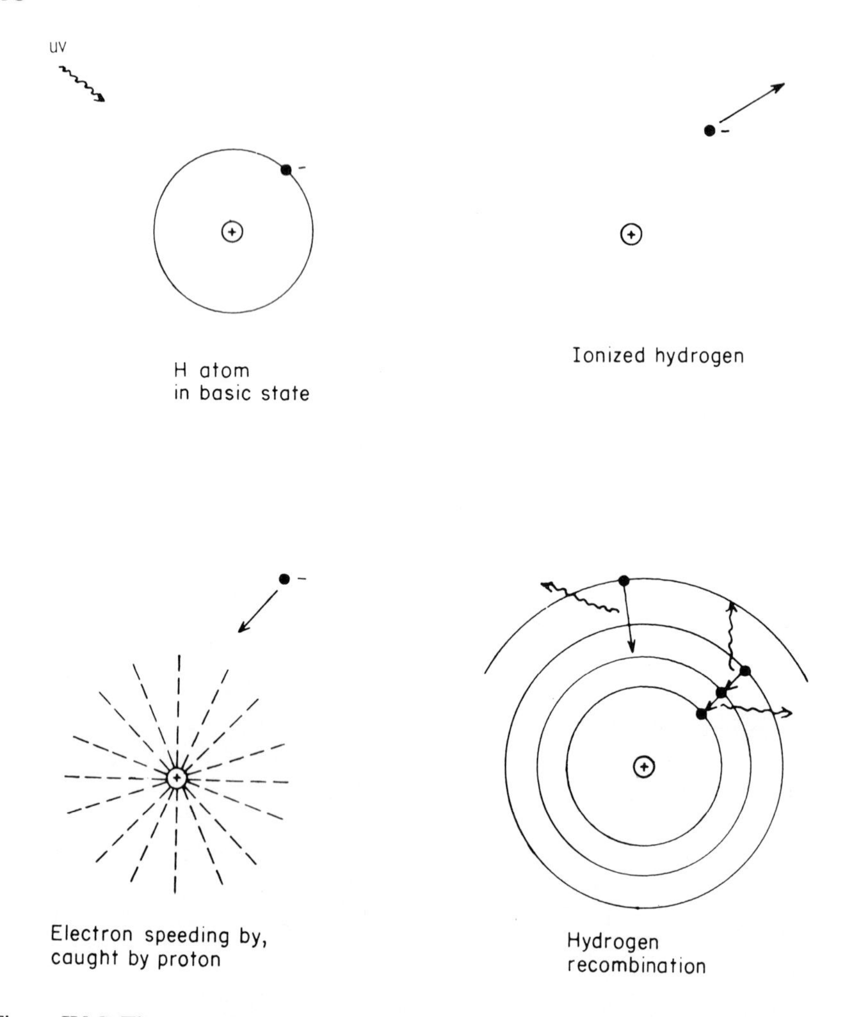

Figure IV.8. The atoms in a gas can be ionized by high-energy radiation and during the process they release one or several electrons. If a free electron comes too close to such an ion, it can be captured and then 'operates' once again within the atom 'union'. If a highly excited atom is caught, then after a time its energy is released in instalments—in the form of light rays. As ionization is carried out through an absorption process but recombination is gradual, the high-energy (= short-wave) light is converted to low-energy (= longer wave) light

special type of interstellar matter which can be seen through a telescope. This is known as the Planetary Nebulae—small, round, bright nebulae surrounding a star. They are called Planetary Nebulae because they appear to be of the disc shape reminiscent of the planets when seen through a small telescope. Just like the diffuse nebulae, they consist essentially of gas lit up by the central stars which they surround. The significant factor is, however, that this gas envelope expands, giving rise to the assumption that it must be created by the star

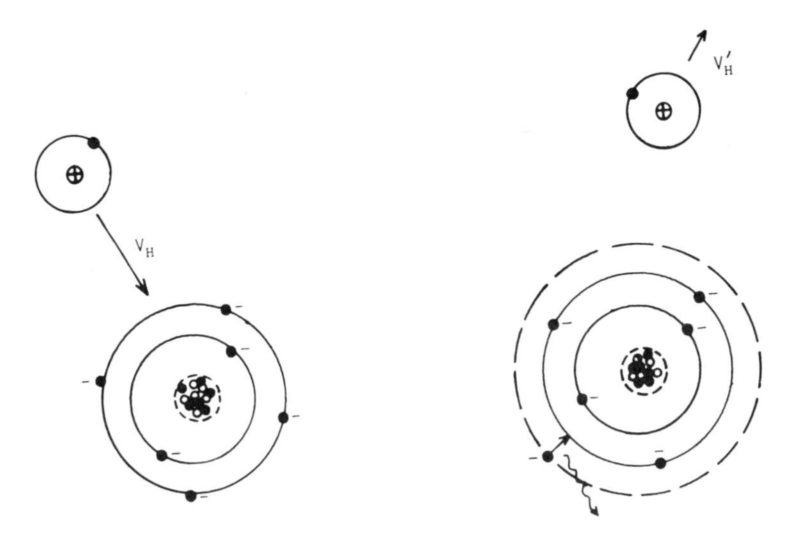

Figure IV.9. Atoms are not stimulated by radiation absorption alone, but also by collision with other particles. In this case, excitation energy is taken from the kinetic energy of the collision partner. If the excited atom releases this energy later in the form of electromagnetic radiation, kinetic energy is then converted to radiation energy

itself—that the star has in fact 'discharged' the matter. This provides us with an important link between stars and interstellar matter; interstellar matter is the raw material for the new star and during the evolutionary period atomic reactions take place, enriching the matter with heavy elements, and in this modified form it is then discharged again into interstellar space at the end of the star's life.

Electronic Ears Hear the 'Void'

We are constantly surrounded by the 'clatter' of electromagnetic waves. Luckily, Man can extract the visible light waves which meet his requirements and ignore the rest, as all the other waves, infrared and ultraviolet light have no relevance to his daily life in general. However, by using modern electronic equipment such as radio and television, even artificially produced waves can be isolated from the electromagnetic field. Thus, radio and television have become 'the small man's electronic ears'. They are not particularly sensitive and therefore are only able to hear the electromagnetic 'din' in the long wavelength range. Over the last 25 years astronomers have developed an extremely sensitive electronic ear in the form of radio telescope which relays 'radiosphere music' from the whole of space. Are the stars the source of this radio radiation? How does the transmitting system work? These were the questions astronomers had to try and answer, and although some of the conclusions were not entirely satisfactory, nevertheless, the results represented a great step forward.

Figure IV.10. The ring nebula in the Lyra constellation is a Planetary Nebula. As the stars evolved, the gas envelope was forced away and is still expanding. The central star provides the light stimulation for the gas. Balmer hydrogen lines and some oxygen and nitrogen lines in particular stand out in the nebula spectrum. The diameter of the Planetary Nebula is several tens of thousands times the diameter of the Earth's orbit and the total mass is 0.2 solar masses. The central star has an apparent magnitude of 15^m. It was called Planetary Nebula because when seen through a small telescope it seemed to be disc, i.e. planet-shaped. (Photograph: Lick Observatory)

One of the first revelations of radioastronomy was that the Sun is also a radio transmitter, but only on a very minor scale in comparison with its light radiation capacity. It seems, therefore, that cosmic radio emission does not come from the normal stars. Interstellar matter, on the other hand, does seem to be a powerful source of radio transmission, surpassing every other atomic and molecular radio transmitter in space. The most significant transmission mechanisms are described below.

1. The 21 cm Line of Atomic Hydrogen

The hydrogen atom is made up of a proton which forms the nucleus and an electron which moves within the electric field of the proton. Both proton and electron have their own angular momentum, or spin; put more simply, both

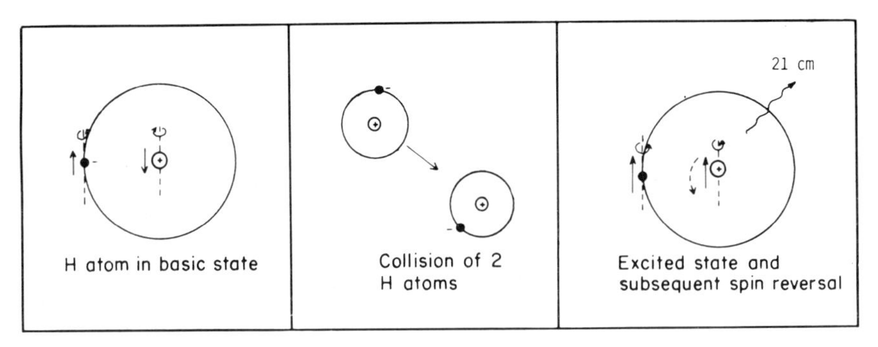

Figure IV.11. When the hydrogen atom is in its original state, the electron and proton spin in opposite directions. However, collisions can alter this situation, giving rise to parallel spin. This involves a higher energy level and excitation energy is again drawn from the impact process. If the spin reverses spontaneously and the hydrogen atom reverts to its original state, an electromagnetic wave is emitted on the 21 cm wavelength. This can be received by radio telescopes

particles rotate. The whole hydrogen atom system can, therefore, adopt specific levels of energy. At the basic, lowest energy level, the proton and electron spins are antiparallel—the particles rotate in the opposite direction. Thus, the hydrogen atom may become so totally disordered on collision with other atoms that the electron spin collapses and then both proton and electron continue to rotate in the same direction. This collapse is accompanied by conversion to a higher energy level, and the energy difference is made up by the particle actually causing the collision. As all atomic systems strive for this basic level, after some time—and in this case we are talking in terms of 10 million years—the electron spin collapses spontaneously and the hydrogen atom emits the energy difference in the form of an electromagnetic wave of wavelength 21 cm. In spite of this exceptionally long waiting period, there are a sufficient number of atoms collapsing at any one time in the Universe, which means that the 21 cm wave is relatively well received and seen on Earth.

2. Radio Recombination Lines

When the atoms which make up an interstellar gas are exposed to high-energy radiation, e.g. UV radiation around a hot star, the atoms become ionized. Atomic nuclei and released electrons then whirl around at random. Now and then, nuclei or ions and electrons come so close to each other that the electron is trapped once more—hence the term recombination. If this takes place at a very high atomic energy level, then the atom will proceed to move towards the basic level in stages ('cascading drop'). At each stage, the energy difference between the two levels transmits an electromagnetic wave. However, the waves emitted during the transition from one high energy level to the next (principal quantum number $n > 50$) are radio waves and are known as recombination lines. Interstellar recombination lines are radiated by hydrogen and helium in and around the edges of H II regions and by carbon.

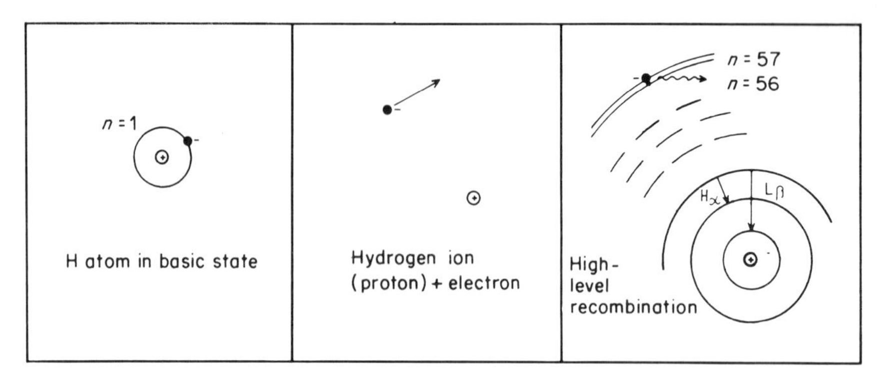

Figure IV.12. As the density in interstellar space is so low, atoms do not collide as frequently as under normal Earthly conditions. Thus, atoms can be very highly stimulated and the electrons move around the nucleus in very large orbits—giving the atoms the chance to 'blow up'. These 'blown up' atoms occur for a short time during recombination in interstellar space. With such a large orbit an ion can trap an electron as it flies past. As the atom reverses to its original stage, the electron jumps from one orbit to the other. The energy difference between two orbits is emitted as an electromagnetic wave. With 'blown up' atoms, these energy differences are relatively small and a low-energy radio wave is sent out. (The principal quantum number for these atoms is greater than 50 and the atoms can grow to about 0.002 mm in size, i.e. about 10,000 times bigger than in the original state)

3. Molecule Lines

Several years ago, radioastronomers discovered sharply defined radio lines (spectral lines in the radio range) in several areas of the sky, especially around the Sagittarius constellation and in the Orion Nebula, lines which had exactly the same wavelength as the radio lines of specific molecules measured in the laboratory. The obvious conclusion was that there must be many molecules in the interstellar cloud which transmit on their own specific wavelength

Molecules are atomic systems with varying degrees of freedom. Just like atoms, they can assume different energy levels, which in their case are a function of the electron arrangement in mutually bonded atoms, of the reciprocal oscillation of these atoms and of molecular rotation. When the rotation level of the molecule changes, a radio wave is generally either absorbed or transmitted. In 1937, the first interstellar molecules (CN and CH) were discovered from spectral lines in ultraviolet light, and then during the 1960s a whole series of molecules were discovered in the radio field. The first of these was hydroxyl (OH) at a wavelength of 18 cm, and this was followed by the more complex ones such as ammonia (NH_3) and water H_2O). By 1975, 59 types of molecule had been isolated in space, many of them organic. The most widely dispersed and hence most significant compounds are carbon monoxide (CO) and formaldehyde (H_2CO). Once formed, a molecule cannot last long in the extreme conditions prevalent in space. If a high-energy light wave from the interstellar radiation field of a cosmic radiation particle hits a molecule, it is then split up again into individual atoms. The life expectancy of molecules

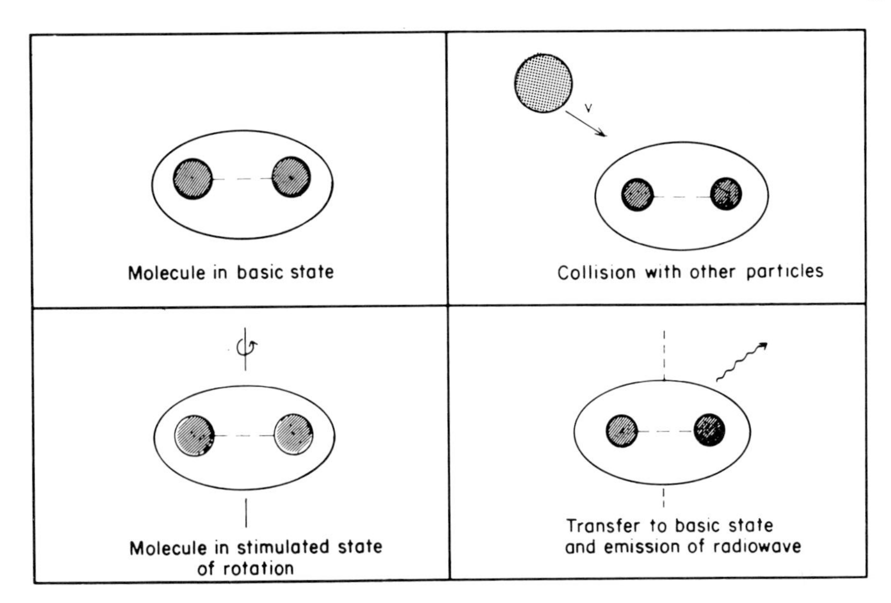

Figure IV.13. Molecules can also emit on radio frequencies. As they are non-spherical in shape, they can rotate around one or several axes. Rotation involves a specific level of energy. However, the molecule is also able to release the absorbed rotation energy as an electromagnetic wave. In interstellar space, any available molecules absorb the rotation energy released during a collision process and then proceed to emit it along radio frequency waves (molecule lines)

actually studied is about 10–100,000 years, and is dependent on type of molecule and interstellar conditions in that locality. For example, molecules in a dense cloud filled with dust particles have a better chance of survival as they are protected by the many dust particles which screen off the interstellar radiation field. In view of their restricted life span, more molecules are being formed all the time. This space chemistry is very different from that carried out in the laboratory and the dust particles are an important contributory factor. Hydrogen molecules (H_2) and probably other compounds can be formed on their surface from hydrogen atoms. Hydrogen molecules and carbon monoxide molecules (CO) are very stable and can act as starting material for other compounds.

Most chemical reactions in space are probably triggered off by high-energy light waves (UV radiation) or cosmic radiation (CR). So, for example:

$$H_2 + CR \rightarrow H_2^+ + e + CR'$$
$$\rightarrow H^+ + H + e + CR'$$

in which H_2^+ is a charged hydrogen molecule from which an electron, e, has been torn away; H^+ is a proton or hydrogen ion and CR′ is the cosmic radiation particle depleted of energy by the reaction. Subsequent reactions may proceed as follows:

$$H^+ + H_2 \rightarrow H_3^+ + H$$

or

$$He^+ + CO \rightarrow C^+ + O + He$$

Complex molecules could then form from these by-products, for example:

$$O^+ + H_2 \rightarrow ?H^+ + H$$
$$OH^+ + H_2 \rightarrow H_2O^+ + H$$

or

$$N^+ + H_2 \rightarrow NH^+ + H$$
$$NH^+ + H_2 \rightarrow NH_2{}^+ + H$$
$$NH_2{}^+ + H_2 \rightarrow NH_3{}^+ + H$$

Complex carbon compounds may also be formed, in accordance with the following reactions:

$$C^+ + H_2 \rightarrow CH_2{}^+ + \text{energy}$$
$$CH_2{}^+ + H_2 \rightarrow CH_3{}^+ + H$$
$$CH_3{}^+ + N \rightarrow H_2CH^+ + H$$

Radioastronomy has discovered 8- and 9-atom molecules such as dimethyl ether [$(CH_3)_2O$] and ethanol (C_2H_5OH) in interstellar space over the last few years.

The opportunity of observing molecular lines in the radio field could be regarded as opening up the way to true space chemistry. However, the chief component of interstellar gas is, of course, hydrogen, present in the form of hydrogen molecules (H_2) in the denser areas. The term molecule cloud implies, therefore, that a considerable part of the hydrogen is molecular and that other molecules are much more in evidence than in normal clouds. In many areas of the heavens, the molecule lines observed for hydroxyl (OH) and water have another special feature—the spectral lines are very intense and extremely narrow. In a normal interstellar gas cloud, the intensity and width of a spectral line is determined by the number of mini-transmitters (atoms or molecules) and by temperature. The more intense lines are also generally broader. However, the molecule lines of hydroxyl and water are sometimes completely the reverse. There must, therefore, be special physical conditions in force in these interstellar clouds. Evidently, the excitation conditions for molecules are similar there to those in many microwave amplifiers, in which suitable molecules are excited and are then converted spontaneously to a level characterized by a particularly long waiting period (metastable level), so that the number of molecules in this state of excitation becomes very high. If an electromagnetic wave of suitable frequency then reacts with the waiting molecules, they become converted to a lower energy level and thus emit a similar wave (induced emision). This can act as a stimulus to others, so that the number of emitting molecules is cumulative and the intensity of the original electromagnetic wave is amplified. This type of amplifier is called a MASER (*M*icrowave *A*mplifier by *S*timulated *E*mission of *R*adiation—known as LASER when applied to the optical light field).

The structure of the OH, H_2O and also SiO molecules is such that they can work on the same principle as a MASER under suitable conditions in space, if they are 'pumped' up to suitably high energy levels. Infrared or ultraviolet radiation can serve as pump energy or an encounter with hydrogen molecules can 'push' the OH and H_2O molecules up to the higher energy levels.

4. Free–Free Transition

Apart from the radio lines described so far, there is also a continuous spectrum in the radio field. If an electron, which carries an electric charge, moves rapidly through an electric field, an electromagnetic wave is emitted. However, in an H II area, the electrons move in the proton electric field and a radio wave is then emitted. This continuous radio spectrum is dependent on particle speed and hence on temperature.

5. Synchrotron Radiation or Magnetobremsstrahlung

Any mobile electron, or more generally any mobile electric charge, is deflected in a magnetic field. This deflection represents an accelerated movement and is connected with electromagnetic wave radiation. Therefore, in interstellar space, the electrons along the lines of force in the interstellar and sometimes stellar magnetic field can be 'pushed along' and mostly radio waves are emitted. This type of radiation is of special importance particularly in Supernova remnants. The first tests were made on particle activators (synchrotrons), hence the term synchrotron radiation.

Thus, accurate recording of the radio waves emitted gives information on the various types of process taking place in interstellar space. The radio continuous spectrum gives information on the movement of electrons in the magnetic field and the temperature and density in H II areas. The chemical composition of many interstellar clouds can be derived from the recombination or molecular lines and the 21 cm line can define the quantity, movement and extended distribution of interstellar hydrogen. Hence the great advantage of radioastronomical observation over optical methods is that it enables us to 'see' much further. Optical methods are almost always clouded to a greater or lesser degree by dust grains, whereas the comparatively long radio waves pass through relatively intact. In this way, radioastronomy has made a considerable contribution to our understanding of interstellar matter. It has shown that the space between the stars is filled mostly with material matter—essentially hydrogen and helium gas, with a much smaller amount of heavier elements and molecules and small grains of dust. This interstellar matter is not uniformly distributed, but cooler, denser clouds 'swim around' in a very thin, hot and relatively evenly distributed gas. The finely distributed gas or stratum and the denser clouds emit on the 21 cm wave. Stars can be formed as the dense clouds contract, but molecules probably accumulate in the concentrated cloud before this stage is reached. However, if a star is actually formed, the raw material

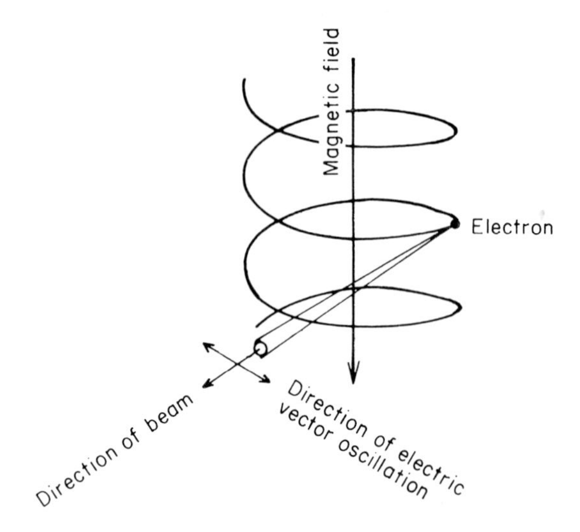

Figure IV.14. If an electrically charged particle is bombarded into a magnetic field, it winds itself along the field lines. As this is an accelerated movement, electromagnetic radiation is emitted. This encompasses a very wide wavelength range (continuous spectrum) and the radiation is polarized. At the same time it is beamed in the direction of the electron path. In the case of one electron alone, the observer would see flashes of light occurring when the beam just happened to be along the line of vision. But when many electrons are involved, these flashes of light overlap and result in a continuous emission of light

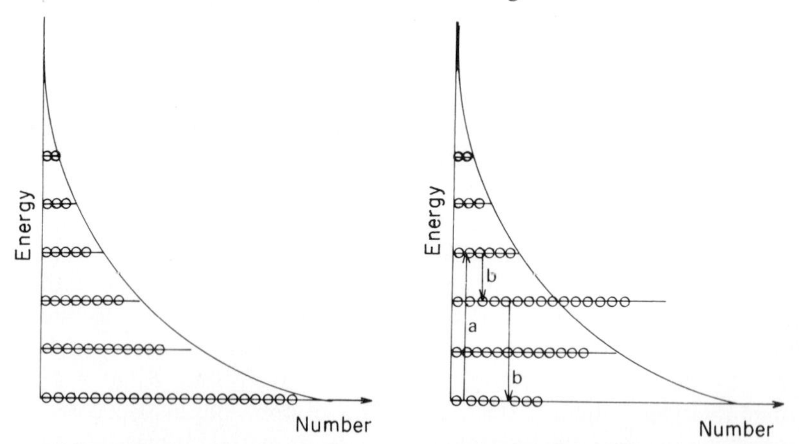

Figure IV.15. In a state of thermodynamic equilibrium there are more atoms with lower energy levels than the higher ones. Excitation and de-excitation processes are balanced out. However, if many particles are stimulated to a metastable level (where certain conversions lower energy levels are then excluded) by pumping (a), they accumulate at this level and there is a reversal of occupation. Radiation of suitable frequency can then induce emission allowing the forbidden conversion to take place (b), and thus intensifying the incident radiation. If light is involved, the mechanism is known as a LASER (*L*ight *A*mplification by *S*timulated *E*mission of *R*adiation) and for microwave radiation it is a MASER (*M*icrowave *A*mplification by *S*timulated *E*mission of *R*adiation)

residue becomes excited, illuminates the star (H II area) and emits on the recombination lines. Old and condemned stars frequently 'discharge' part of their mass. Even this waste material is illuminated (Planetary Nebula and Supernova remnants). Clouds with a particularly high dust grain content can be seen around stars by reflected light. Other similar clouds absorb so much starlight that they appear as holes or gaps in the stars.

The density of the various phenomena connected with interstellar matter covers a very wide range. A figure of 0.1 particles/cm^3 is adopted for the general stratum or intermediate cloud gas, whereas normal clouds are between 100 and 1000 times more dense and up to 10000 particles/cm^3 can be found in the middle of diffuse nebulae such as that around Orion. Even greater densities have been found in molecule clouds which are made up essentially of H_2 molecules plus some rather 'exotic' components. Their density is between 10^3 and 10^7 particles/cm^3 on average. In comparison with our test laboratories on Earth, this still represents an ideal vacuum. Even the most 'empty' laboratory chambers contain almost 10^9 particles/cm^3, more than even the very densest molecule cloud. In view of this, an experimental physicist would be right in saying that radioastronomers actually hear the 'void'.

V

The star population

The heavens are populated with numerous stars and interstellar matter. The stars themselves have always been regarded as individuals, whereas, in fact, many of them are 'Siamese twins or triplets' due to gravitationally bound double stars and multiple systems.

Just as the population of a town or county is not given just as a list of individuals but is also defined as to number of and size of families, so the stars can be divided up into 'family groups' in a similar way. However, there are outsiders in addition to families and various star communities which differ from each other in many ways. We already know that stars have different spectra and can, therefore, be divided up into spectral groups. However, these classifications are based on 'external' features and so do not represent any family relationships (just like the categorization into 'blonde, brown or black-haired' people says nothing about their family connections). Hence spectral classification is no guide to star relationships. However, there are star clusters which did not occur by chance or were not formed by celestial sphere projection, but which represent genuine, genetically controlled star concentrations in a relatively small area of space.

What is a Star?

Up to now, the concept 'star' has been applied to any bright, luminous point in the night sky, the only differentiation being between the fast moving planets and the apparently immobile fixed stars. We have seen that there are many bright, luminous stars which can be grouped into constellations and thus present a clearer overall view of the heavens, but that there are far more less bright stars. This still reveals nothing about the nature of the stars, but a knowledge of both this aspect and their structures is essential if some sort of 'order' is to be made of the heavens.

The astrophysical definition of a star is that a star is a self-illuminating gas sphere, held together by its own gravitation. Definitions are generally short and precise and they are, of course, necessary, but this particular definition is not sufficient for our purposes, as we need to investigate the stars in more detail. Let us begin by assuming that our Sun is a completely normal star, as to be found many thousands and millions of times in space. From its external appearance, the Sun seems to be a brightly lit sphere of about 1.4×10^6 km in diameter and a mass of 2×10^{27} tonnes, or 340,000 times that of the Earth (obviously the Sun cannot be weighed, but the mass can be calculated from the gravitational pull between the Sun and Earth). Spectral analysis of the Sun's light shows that this incandescent gas sphere consists predominantly of about 65% hydrogen and 32% helium plus a small quantity of heavy elements.

Just as any terrestrial object is attracted by the Earth and thus endowed with a certain weight, so mass particles gravitate towards each other on the Sun. This gravitational force attempts to draw all particles to the centre and, in principle, this does seem possible, as gas particles are suspended and a gas can be compressed. However, a second force prevents the accumulation of all particles in the central area, a force which is anti-gravitational in effect and is derived from the gas pressure forcing outwards all around and which increases as the gas temperature rises. In a stable star—and our Sun is one of these—the two forces are in equilibrium, i.e. they balance each other out. However, if a star were to be compressed further by an imaginary 'giant's fist', then its internal temperature would rise and this would in turn cause an increase in gas pressure. This pressure would then overcome the gravitational force and the star would expand and thus regain its original state.

One very important property of the stars is their light and heat emission. As both light and heat are special forms of energy, there must be an 'energy dispenser' somewhere in the star so that it can radiate and remain stable at the same time. If there were no such energy-releasing mechanism then the star would lose its internally stored energy as light was emitted, causing a reduction in temperature and a disturbance in the pressure/gravitation balance in favour of the latter. The star would then slowly contract. This contraction, during which gravitation energy is converted to light and heat, could be a form of energy itself at certain periods of the star's development. However, estimates showed that this form of energy release was not sufficient for normal stars like our Sun, as the contraction energy was not powerful enough to cause the Sun to shine in the way it does and has been doing for many hundreds of millions of years.

Astronomers therefore had to look for other energy sources to explain star radiation and much information has been obtained this century from the realm of atomic physics. During the course of laboratory tests on the atom, it was found that the atom nucleus changed when bombarded with a rapid stream of hydrogen or helium nuclei; it either absorbed them or was split into several parts. There was also a possibility that helium nuclei could be constructed from several hydrogen nuclei. The special feature of this nuclear structure is that the individual sections added up together weigh more than the finished nucleus itself. However, the 'lost' mass does not vanish without trace, it is converted to energy, and to radiation energy in particular. The conversion of hydrogen to helium nuclei, so-called nuclear fusion, seemed to be an extremely abundant source of energy for stars, as the hydrogen atoms deep inside the star move so rapidly because of the high prevailing temperature that they act as projectiles, and both targets and projectiles are plentiful as a result of high pressure and high material density. The energy released during the fusion process is given off mainly in the form of γ-rays. On its way from the star centre to the star surface, this radiation is absorbed by very many particles and radiated once again. The rays gradually lose their energy (but the total energy remains unchanged), γ-rays become X-rays, then UV light, and finally leave the stars as visible light.

Although we are unable to see into the stars and so cannot actually examine these processes direct, calculations based on this principle agree well with observations made, so there seems little doubt that the concept is basically correct. The calculations represent a type of X-ray vision into the stars. Let us know consider the Sun in the same way. Density and temperature are greatest in the centre; the former is about 130 g/cm^3 and the latter about 15×10^6 K. Although the density is many times greater than that of any metal, the material reacts as gas because of the extreme temperature conditions. However, in contrast to normal gas used in the laboratory, the gas particles in the Sun are not electrically neutral; they have lost most of their electron shell and are fully ionized. It is a gas composed of naked atom nuclei, in which collision is a common occurrence and some atomic fusion as described above takes place with attendant release of energy. Temperature, density and pressure are reduced towards the outside. The limit of energy release is reached at a distance from the centre corresponding to one fifth of the Sun's radius. Areas outside this limit do not contribute to the star's luminosity, they merely act as passageways through which the radiated light travels. The very short wavelength rays zig-zag through; they cover a distance of about 2.5 cm between each 'zig' and 'zag' and are first of all absorbed and then re-emitted, and the further they stretch towards the surface of the star, the longer the wavelength becomes. The actual surface of the star forms a thin (at least compared with star size), hazy atmosphere, in which pressure and density drop rapidly and become like that in the rest of space. The atmosphere outside the corona is about 10,000 km thick and the actual radiation layer, the photosphere, is only 300–400 km. Both the photosphere and the chromosphere above it form the actual star face and it is here that the radiation acquires all its characteristic properties.

The lower photosphere of a star emits radiation in which all wavelengths are superimposed (continuous spectrum), but maximum intensity may occur at different levels. The higher the temperature of the photosphere, the shorter the wavelength at which maximum intensity is reached. This continuous radiation is represented by the characteristic absorption line spectrum of the photosphere and an emission line spectrum from the chromosphere. The type and thickness of the spectral lines depend on the type and number of atoms present, their ionization level and thus on the temperature and pressure in the photosphere and chromosphere. Accurate analysis of the spectrum can,

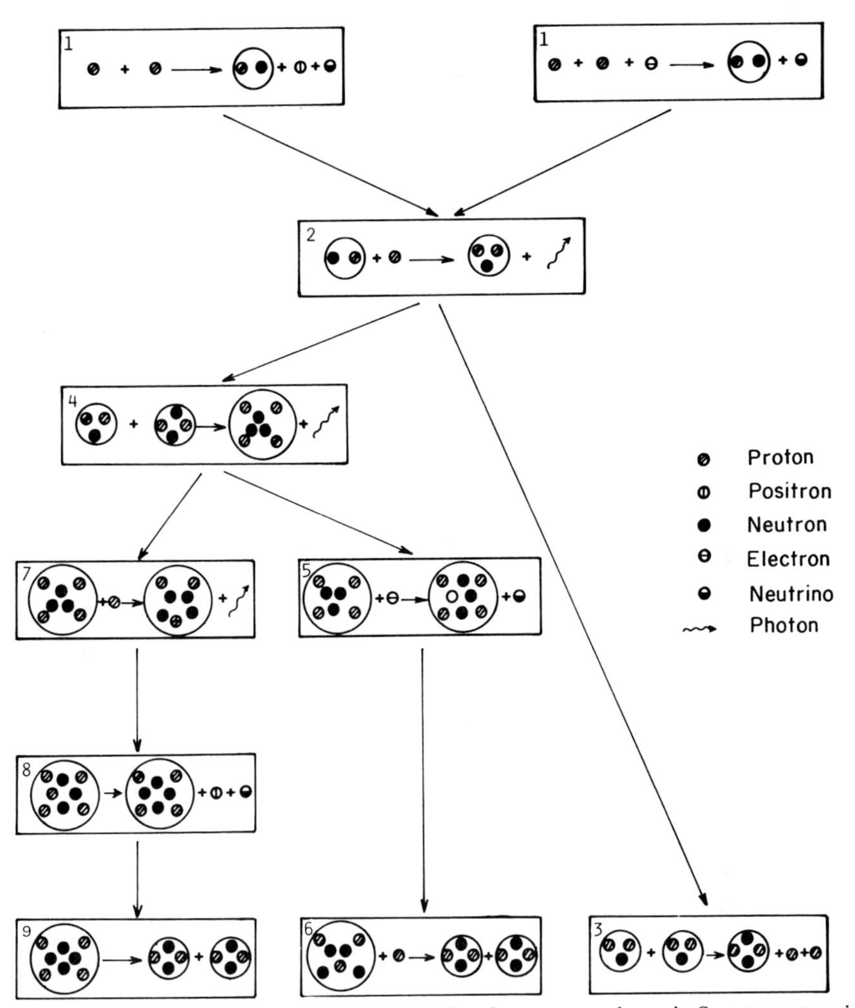

Figure V.1. The most important nuclear reaction for energy release in Sun-type stars is the proton–proton chain. In more than 99% of cases, two protons react together to form a deuterium nucleus, with simultaneous emission of a neutrino and a positron. But sometimes two protons and an electron combine to form one deuterium atom and one neutrino (1). By absorbing a further proton, the deuterium nucleus becomes converted to a helium-3 nucleus and a photon is emitted (2). About 86% of helium-3 atoms then undergo an exchange effect; one helium-4 nucleus and two free protons emerge from two helium-3 nuclei (3). On the other hand, one helium-3 nucleus plus one helium-4 nucleus gives beryllium-7 and one photon (4). The beryllium-7 can then be 'processed' in two ways: beryllium-7 and an electron give lithium-7 and a neutrino (5). The former combines further with a proton to give two helium-4 nuclei. The second possibility is that beryllium-7 absorbs a proton to give boron-8 and a photon (7). The boron-8 nucleus is not stable and decomposes into one beryllium-8 nucleus, a positron and a neutrino (8). The beryllium-8 nucleus also is not stable and it proceeds to decompose to two helium-4 nuclei (9). In each case, the reaction chain has formed a helium nucleus from four protons, but the helium nucleus has a lower mass than the four protons. The mass difference is converted to energy during the reaction period and the Sun then emits this in the form of photons

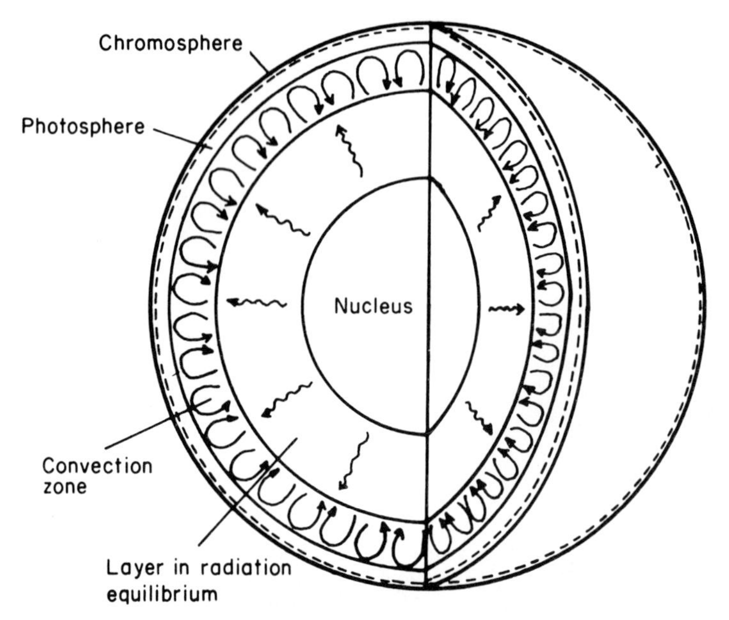

Figure V.2. In the Sun's nucleus, energy is released by the proton–proton chain and transported away in the form of radiation (atoms absorb, re-emit and scatter the radiation). Further out, the Sun's matter is not so well structured and is subject to continual up and down movement. In this convection zone, energy is transported by hot matter rising and thus cooling and the cooled matter falling to be heated up again. Above this layer there is the Sun's atmosphere with its 400 km thick photosphere and the chromosphere above that, marking the beginning of space. The chromosphere (= colour envelope) gets its name from the fact that it appears as a red-glowing layer during a total solar eclipse

therefore, give important information about a star such as the pressure and temperature of the outer layers. The pressure in the star's atmosphere is influenced by the prevalent acceleration due to gravity and this can also be found. Then again, the diamater of the star can be calculated from temperature measurements in conjunction with total energy released by the star (this is, in fact, the luminosity). Finally, star radius combined with acceleration due to gravity allows calculation of star mass.

Many stars have been examined in this way and their variables of state defined. One astonishing feature emerging from the results is the wide luminosity range of stars—from one ten thousandth to one million times the Sun's luminosity—over an extremely small mass distribution range; the stars with the lowest mass amount to one twentieth of the Sun's mass and those with the highest are 50–100 times greater than the Sun's mass. Radius and temperature may also vary considerably. The smallest visible stars have a radius of only one hundredth the size of the Sun's radius (only slightly more than that of the Earth), whereas that of the largest stars is almost a thousand times greater than that of the Sun: the whole orbit of Mars could be contained within such gigantic stars.

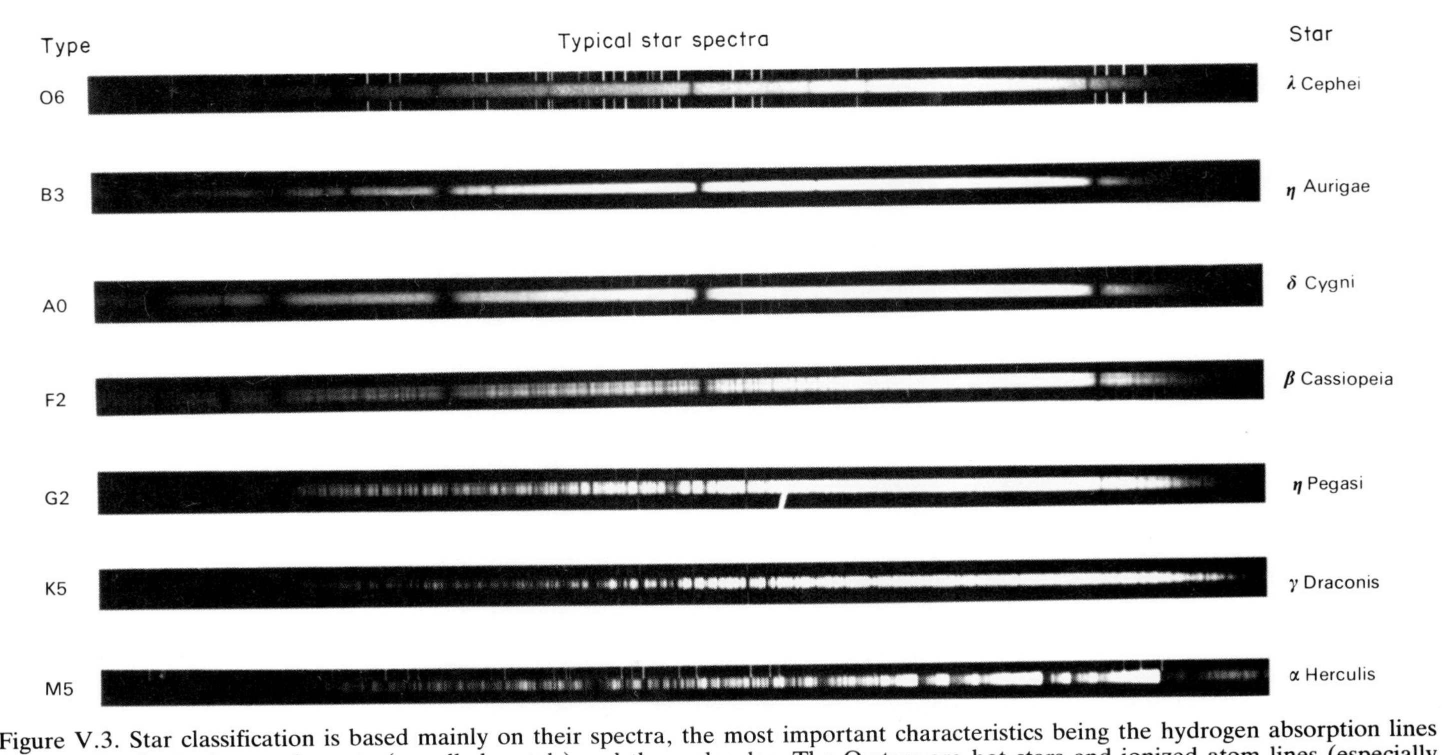

Figure V.3. Star classification is based mainly on their spectra, the most important characteristics being the hydrogen absorption lines (Balmer lines), the heavy elements (so-called metals) and the molecules. The O stars are hot stars and ionized atom lines (especially helium) appear in their spectra. Balmer lines are faint. Sometimes emission lines appear. With the B stars, neutral helium lines predominate and Balmer lines are stronger. The latter attain maximum intensity in the A stars, from which lines of ionized heavy elements (e.g. calcium) are also seen. The calcium lines are particularly strong in the middle spectral types F and G (left in the spectrum). G star spectra are similar to that of the Sun (the Sun itself is a G2 star). The later K and M stars have numerous, intense metal lines and molecular bands are visible in longer spectral wavelength areas (right); that of titanium oxide (TiO) is especially impressive. The individual spectral groups are then subdivided (e.g. A0, A1, A2, ... , A9). The picture shows the star spectra with the typical dark absorption lines. Above and below is the orientation spectrum with the bright emission lines, which is often produced by an iron arc. The orientation spectrum serves to identify the stellar spectral lines and establish a wavelength scale. (Photograph: Mt. Palomar Observatory)

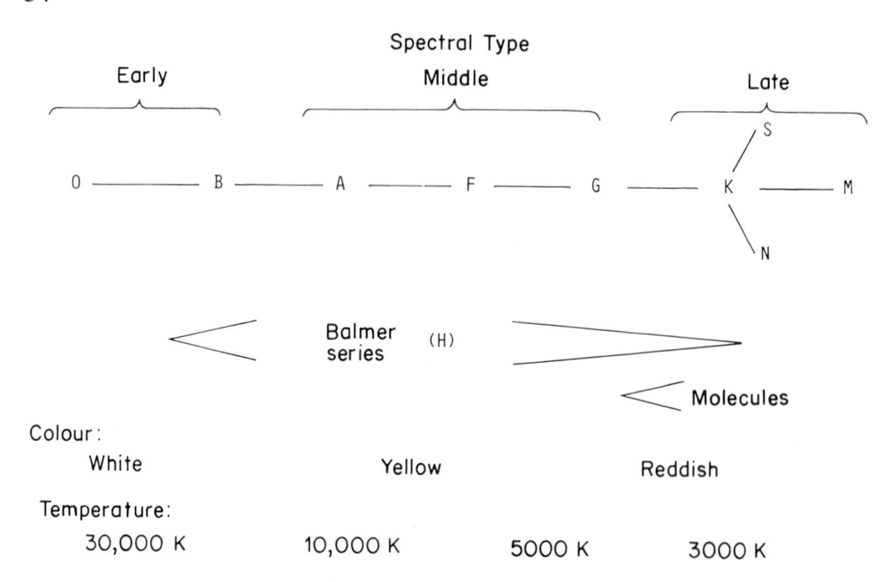

Figure V.4. The star spectra are classified according to various criteria and the categorization also covers temperature and colour sequence. Several emission lines can be seen in the early spectral types O and B. These then disappear and absorption lines become more frequent. The hydrogen lines, particularly the Balmer lines in which the hydrogen atom falls from a high excitation level to the second one, are of special importance. In the early spectral types, the Balmer lines are weak, they become particularly strong in the A type and then weaker again in the later F and G types. Many neutral atom (e.g. iron and calcium) and molecule band lines (especially those of TiO) are found in the K and M types

The different variables of state for a star are not completely independent of each other. For example, if the luminosity or absolute brightness of very many stars is plotted against spectral type (which is closely connected to surface temperature), then a distinct pattern emerges. The Danish astronomer, Ejnar Hertzsprung and the American Henry Norris Russell were the first to discover this and the resultant graph (Figure V.5) is named after them. The numerous main sequence stars, the 'normal citizens' of the star population, form a diagonal pattern. The large, luminous, so-called Red Giants are found above. The difference between these and the main sequence stars is that their temperature range is relatively lower at the same degree of luminosity, owing to a particularly high surface area. Although they do not have the large mass which comparable main sequence stars have, they do have a large radius—they are, as it were, inflated. However, they also differ from normal stars internally; in the main sequence of stars, energy is released in the central section by the fusion of hydrogen atoms to give helium nuclei, but the giant stars have a spherical shell in which helium is synthesized and in the central area carbon and oxygen are formed from the helium 'ash' resulting from hydrogen combustion.

The so-called White Dwarfs, featured on the bottom left of the Hertz sprung–Russell diagram, are even more exotic than the Red Giants. They have

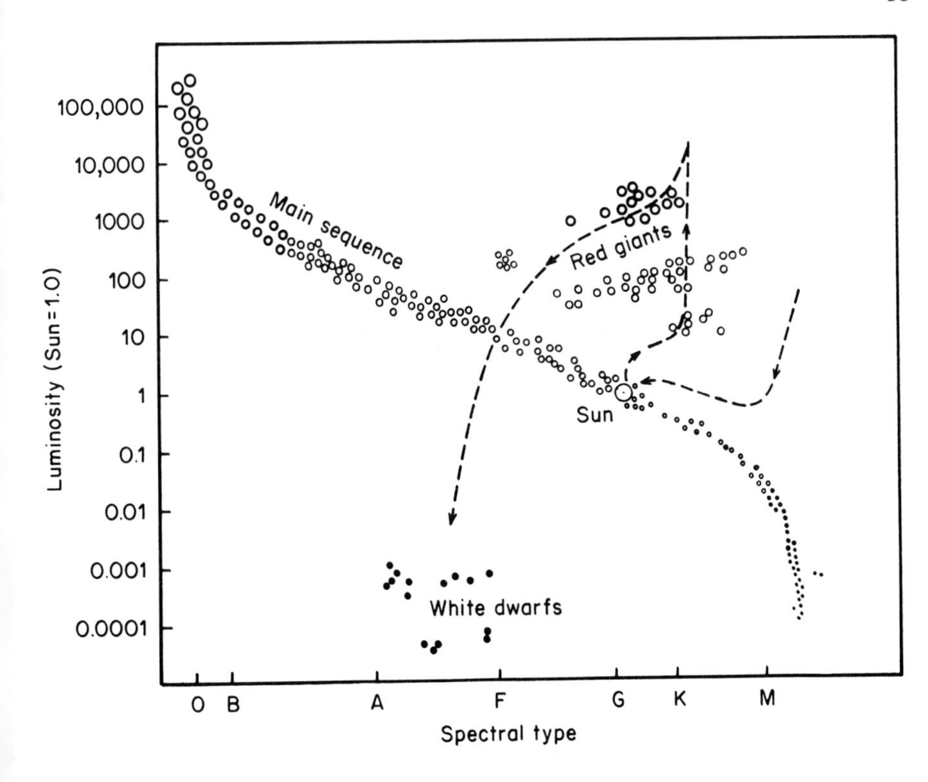

Figure V.5. In the Hertzsprung–Russell diagram, spectral type is plotted along the abscissa and absolute magnitude or luminosity along the ordinate. If the data from very many stars are plotted on such a diagram, three densely populated areas emerge: the main sequence, the Red Giant and the White Dwarf areas. Star mass and radius increase along the main sequence from bottom right to top left. This mass–luminosity and mass–radius relationship applies only to main sequence stars. As the stars evolve, size, spectral type and absolute brightness all change as a function of time and the star 'wanders' across the diagram. The dashed line represents the Sun's evolutionary path

a mass approaching that of the Sun, but are only as large as the Earth, which means that the average internal density must be very high: 1 cm^3 of White Dwarf matter would weigh approximately 1 tonne. Moreover, very many of them are probably made of iron, so no more energy is released internally and they cool down much more slowly. This statement seems to contradict the above assertion that stable stars—and the White Dwarfs do fall into this category—must contract without energy release. The White Dwarfs have been through this contraction process and the matter contained in them is compressed to the maximum under prevailing physical conditions. Now the electron pressure is the only counteractive force of gravitation, but this electron pressure is independent of temperature. As these stars are so small and have a correspondingly small surface area, their luminosity is also very poor and they are difficult to detect in the sky. The first star which was accepted

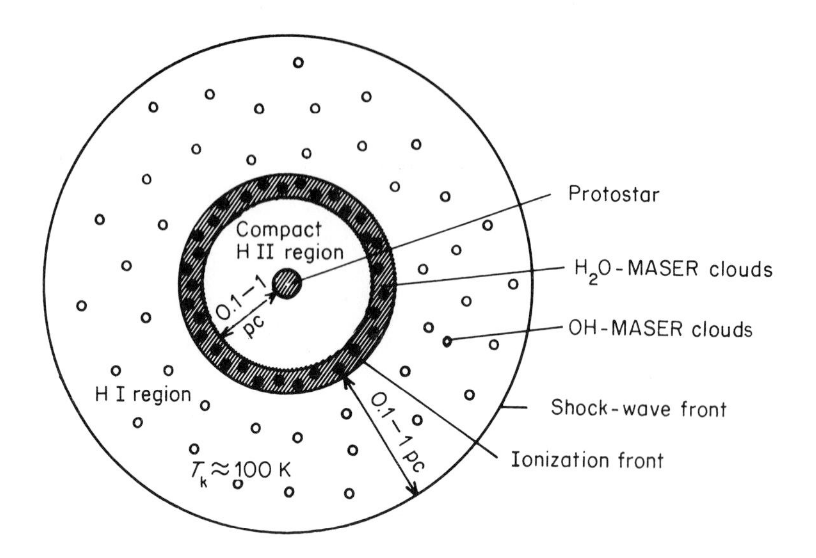

Figure V.6. The protostar is probably surrounded by a compact H II region enclosed by the ionization front. Smaller H_2O-MASER clouds could have formed near the ionization front. A shock wave has passed through the outer H I region behind which OH-MASER clouds could be formed. The relatively high dust density makes the protostar optically invisible and only the radio radiation from the MASER clouds gets through to us

into the White Dwarf category was the companion of the brightest star in our sky, Sirius.

Now that we have learned about different stars, we may find it useful to know whether they are related to each other in any way and how the internal 'nuclear oven' can affect changes in the star. To this end, a short star biography now follows!

Under certain conditions, a cloud of interstellar matter can contract through its own gravitational force. Should this occur, all the free-fall particles gravitate towards the centre, thus increasing the density at that point. As long as the cloud remains transparent to allow the radiation released to pass through (converted gravitational energy), the temperature remains unchanged. However, as density increases further, the innermost region becomes non-transparent, the released energy cannot be emitted and there is a consequent rise in temperature. Contraction stops and further cloud matter 'rains down' on to this central area from outside. Temperaure continues to rise until the hydrogen molecules are dissociated. This latter process consumes energy and can, therefore, take over the role originally played by energy radiation, so that the central area can start to contract again. When all the hydrogen molecules are dissociated and later on even the hydrogen atoms in the centre are ionized, the central temperature rises again, contraction starts to slow down and finally comes to a halt. At a temperature of about 500,000 K, the first nucelar fusion processes begin, in which lithium beryllium, boron and deuterium are 'burned'· If the central temperature increases to several million degrees at a density of 100 g/cm^3, hydrogen nuclei begin to fuse together and a star has been born. The central region of a contracting cloud is known as the 'protostar' during the phase before hydrogen combustion starts. As a rule, the protostar is not directly visible; as the cloud matter contracts, the interstellar particle density increases and probably forms a sort of 'cocoon' which hides the emergent star.

The protostar radiates energy at a relatively early stage; the dust grains inside the cocoon start to fuse and vaporize, contraction in the external areas of the cloud stops and the matter is ionized. A compact H II area has formed around the protostar. The boundary between ionized and neutral matter (ionization front) is well defined and becomes rapidly 'eaten away' towards the exterior until geometric radiation dilution (the same amount of radiation takes up an every increasing amount of space) and absorption by recombined hydrogen atoms slows down the rate of advance and the ionization front moves more slowly than the sound velocity in the gas. Ionization provides two particles from one atom and both particles consume additional energy; thus, both pressure and temperature rise in the H II zone and the gas expands with sound velocity, forming a pressure front. When the ionization front slows down, the pressure front catches up and then even overtakes it. However, as the sound velocity in the ionized gas is much greater than that in the neutral gas, after overtaking, the pressure front runs into the neutral gas with ultrasonic velocity. This type of process is called a shock wave. Further compression on a lesser scale, in which molecules—especially H_2O molecules—are formed, can occur in the transition zone between the compact H II region and the neutral gas cloud. These lesser condensation processes (10^8–10^9 km) act as an interstellar or circumstellar MASER. There are similar structures to be found further out, but these 'work' with the hydroxyl molecule and are therefore known as OH-MASERs. Circumstellar dust grains prevent us from seeing the protostars, but radio radiation from the MASER passes through intact; the MASER represents an extremely short-lived phenomenon by cosmic standards, so MASER radiation can be regarded as a 'baby's first cry'.

Hydrogen combustion marks the end of the star's birth—it is now in its normal state and is located in the main sequence on the Hertzsprung–Russell diagram. It will remain there for most of its 'life'. This 'life' is longer the smaller is the star's mass. This seems paradoxical at first but emerges from the fact that the high mass stars are much more wasteful with their energy reserves than are the thrifty smaller stars. In time, the helium ash from the star reactor accumulates in the central part of the star. When this ash content has reached a level of about 15% of the total star mass, the helium sphere begins to collapse under its own weight, it contracts and the central temperature and density rise until nuclear fusion is renewed and helium is 'burned'. As the nucleus area is contracting, the star envelope is expanding; it finally blasts forth and becomes a Red Giant. Under suitable conditions, this process of 'ash' contraction and subsequent renewal of nuclear fusion can be repeated several times to provide any further source of energy. The star becomes a White Dwarf—although under certain circumstances it may need to 'discharge' a little more of its mass before it can fully qualify for this category. However, a star only reaches its 'pension stage' this easily if its mass is not much bigger than that of the Sun. Extremely massive stars suffer catastrophic experiences (e.g. Supernovae), lose a considerable part of their mass and either become neutron stars or are totally ripped apart, or alternatively, contraction can continue until a 'Black Hole' emerges. (A neutron star consists predominantly of neutrons or neutron-enriched atomic nuclei, whereby the neutrons are non-electrically charged fundamental particles and have the same mass as protons. The neutron star mass is similar to that of the Sun in size, but its diameter is only 10 km, giving a density in the region of 10^8 tonnes/cm^3. As this corresponds roughly to the density of the atomic nucleus, for the sake of simplicity the neutron star can be regarded as a 'gigantic atomic nucleus'. However, in the Black Hole, contraction of stellar matter is far advanced, such that the attractive force on the star surface has become so large that no more light can get through. Thus the object becomes invisible to us and merely continues to 'suck in' both matter and light.) White Dwarfs, neutron stars and Black Holes—and the last are still hypothetical—are thus the 'pensioners' of the star population.

Variable and Binary Stars

Apart from seasonal variations, a passing glance at the sky would not register any change— but changes do occur and some of them can be seen with the naked eye. If the Algol star (or β Persei) in the Perseus constellation is studied carefully over a period of a few days, its brightness will be found to alter distinctly. Not just a handful, but thousands of these variable stars have been isolated to date, many have had light change accurately plotted over a period of time and the resultant curves have been divided up into different categories. The most significant group is that of the periodic variable stars which show a cyclic light change—the cycle lasting anything from one hour to several years and the light fluctuating from between several hundredths of a magnitude up to several magnitudes. These variable stars can be subdivided again. Firstly there are the long period variables with a cycle in excess of 100 days and most of them having a very wide amplitude, as represented by the star Mira in the Cetus constellation. The δ Cephei stars form a second category; they have an extremely regular light variation pattern ranging from 1 to 50 days, but the amplitude is smaller than that of the long period variables. Note that there are actually two types of cepheid variables—those present in the globular cluster and those found in the open star cluster and in the star field in general. The latter are the classical cepheid variables. Finally, there are the short period RR-Lyrae stars with cycles lasting about one day. Both cepheid variables and RR-Lyrae stars have a particularly interesting characteristic feature; the RR-Lyrae stars all have about the same degree of luminosity and the cepheid variables show a marked correlation between length cycle and luminosity. As

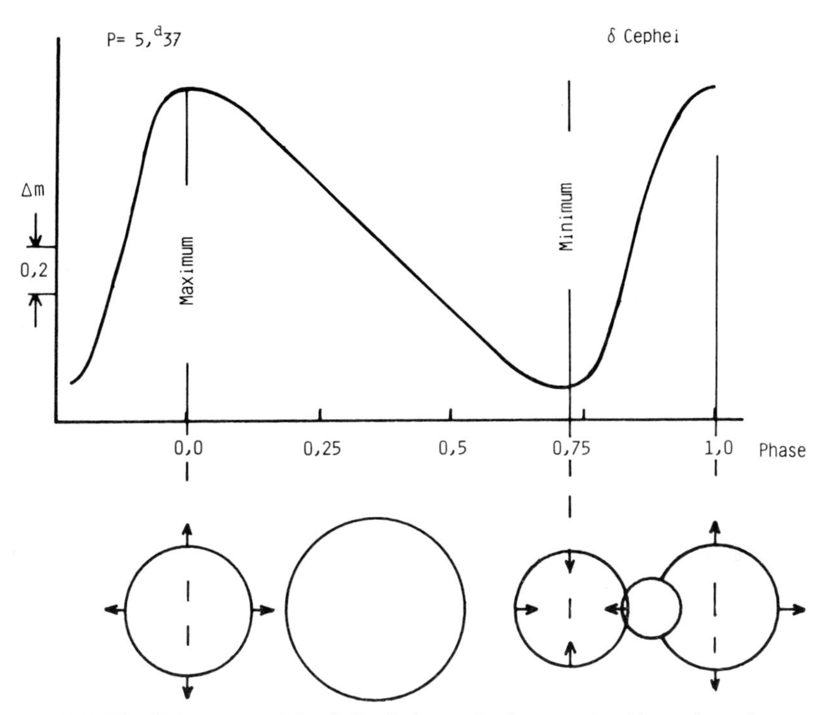

Figure V.7. The light curce of the δ Cephei stars is characterized by a sharp increase in luminosity and a much slower luminosity reduction path. The brightness course is due to a pulsation in the upper star layers. The reason why star diameter does not change in proportion to its brightness is because the external layers cool down during expansion and thus reduce their emission level. The opposition between the two processes—increase in star diameter and hence increase in radiant surface and brightness and cooling of the uppermost layers followed by reduced luminosity—is the reason for the phase difference observed between maximum luminosity and maximum expansion of the star

the variable star cycle is an easy factor to ascertain, it can then be used to work out luminosity and absolute brightness, and photometric methods can subsequently be used to calculate distance. Astronomers now think that they understand something of the mechanism of these variable stars. They seem to expand at certain time intervals and then contract again. This change in size due to expansion and contraction of the upper star layers obviously changes the radiant surface and hence luminosity. As these variable stars 'pulsate' regularly, they are known as pulsation variables. Then there are also the irregular variable stars, the mechanism of which is relatively unknown to date. Could they be 'sick' stars, perhaps? At the beginning of the section, the Algol star in the Perseus constellation was mentioned. When its light curve, which is again strictly cyclic, is drawn up, it looks completely different; the star shines steadily for a long time and then suddenly loses a little of its brightness. This behaviour shows that Algol is not a variable star in the physical sense at all, but an eclipsing binary, i.e. a double star, which has its brighter components

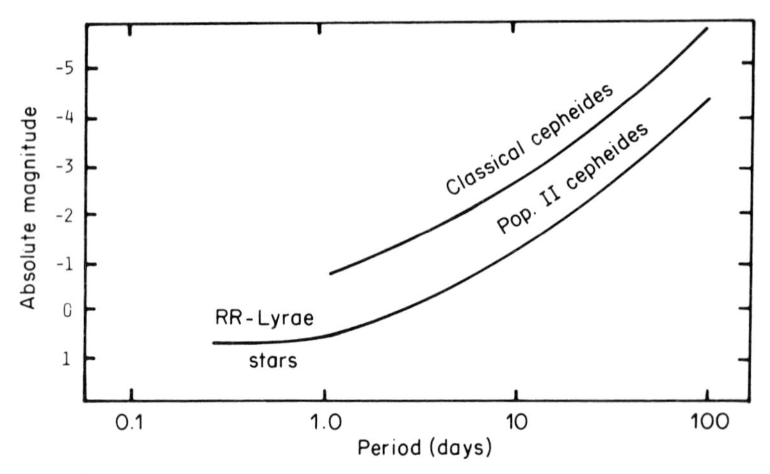

Figure V.8. The period–luminosity relation of RR-Lyrae and δ Cephei stars shows the increase in absolute luminosity of the Cepheides as a function of period. The RR-Lyrae stars, on the other hand, have the same absolute magnitude regardless of time. The classical Cepheides found in the galactic disc are about 1.5 stellar magnitudes brighter than those in population II found in globular clusters. These latter are known as W-Virginis stars

'hidden' from us now and again. Both components of the binary system thus move elliptically around the common centre of gravity. Such close double stars, neither of which can be seen individually, are also revealed by their spectra, as both components alternate between movement towards and then away from us and the Doppler effect causes the spectral lines to shift periodically. However, there are also double stars which are actually recognizable as such through a telescope and the orbits of which can be measured with a micrometer. This type of star bond is called a visual binary. Sometimes a double star will revolve around a single star to give a triple system. Systems of three or more stars are called multiple systems, as a general term, and a catalogue listing stars close to the Sun shows that 59% of all stars are components of double or multiple systems.

Some stars have been of special interest to astronomers over the last few years. When observing the heavens through X-ray telescopes which are incorporated into artificial satellites or rockets to escape the X-ray-absorbing effect of the Earth's atmosphere, they found intensive radiation in the X-ray range, but it was severely reduced from time to time. In view of these periodic changes in intensity, astronomers concluded that these were double stars emitting the compact X-rays, in which one component is a highly luminous expanded star around which the second component, which is an extremely compact star with a material density like that of an atomic nucleus, revolves at a relatively short distance away. As the stars are close together, gas from the expanded luminous star (which is responsible for the optically visible radiation) permanently shrouds its solid companion. This gas coils its way down to the surface of the second star where gravitational energy is then exchanged for kinetic energy. The ionized gas is stopped at the surface; this deceleration is

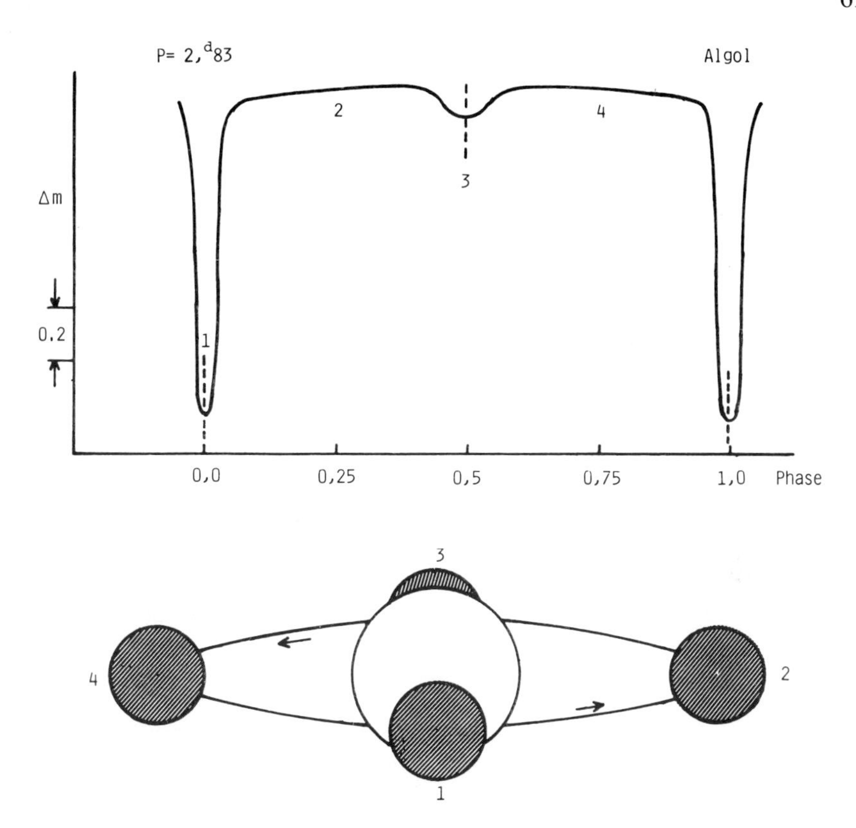

Figure V.9. Light curve of Algol (β Persei). The low minimum values occur when the companion covers the main components and the smaller dip occurs when the companion itself is covered. The companion, which is of low light intensity, and the main star are shown in four relative positions. Positions 2 and 4 represent the constant part with the light curve at maximum luminosity, 1 is where the low minimum occurs and 3 the smaller one

equivalent to a highly negative acceleration of charged particles and results in electromagnetic wave emission. As the energy level of particles is high, the wavelengths radiated are very small and this is, in fact, X-radiation. The X-ray source pulsations are due to the fact that the larger components hide the solid companion from our sight once in every orbit. Most research on X-ray double stars to date has been focused on Cen X-3 (the third X-ray source in the Centaurus constellation), Her X-1 and Cyg X-1. Her X-1 was identified optically with the HZ Her star in the Hercules constellation, which was recognized as a variable star as far back as 1941. This system has a pulsation period of 1.24 s, overlapped by a sinusoidal variation of the period of 41 hours. It can all be explained as a double star system which is made up of a neutron star 1.3 times the Sun's mass and a star 2.3 times the Sun's mass with a diameter of at least 5×10^6 km. The orbital diameter is about 6.2×10^6 km.

Thus, X-ray double stars provide the means for observing directly extinguished stars, White Dwarfs, neutron stars and maybe even Black Holes.

Open Star Clusters

Two accumulations of stars in the Taurus constellation, the Pleiades and the Hyades, can be seen with the naked eye but closer inspection revealed that they are in fact star clusters. Two more spectacular clusters, h and χ Persei in the Perseus constellation, can be discovered through field glasses. The diameter of these open star clusters is between 2 and 20 pc and each contains between 20 and 300 visible stars. In the Pleiades, for example, there are at least 120 stars in a 4 pc diameter sphere, giving a minimum star density of 4 stars/pc^3, which is a very high figure when compared with the 0.05 stars/pc^3 in the vicinity of the Sun. Many open star clusters have an even greater density and are not characterized by concentration alone, but by their general movement too. Even if one or two stars do manage to 'break away', most of the members remain chained together by the force of gravity. This physical unity becomes even more apparent when the star cluster is plotted on a Hertzsprung–Russell diagram. If there is a large enough number of stars, the main sequence stands out clearly. The right-hand boundary of the main sequence is dependent on instrument range, the left-hand side stops dead and several stars are often found in the Red Giant area in the top right-hand section. Comparison between the Hertzsprung–Russell diagram of a star cluster and that following the development of a star described earlier in this book shows that a star cluster contains stars in various stages of their development; these define the evolutionary trend of stars in the Hertzsprung–Russell diagram. The sudden discontinuation on the left-hand side of the main sequence then means nothing more than the fact that the stars with greater mass which develop much more quickly than the lower mass stars, have already 'wandered' from the main sequence to join the Red Giant or maybe even the White Dwarf group. In view of this, it is reasonable to assume that all the stars in a cluster were formed from a large interstellar cloud at about the same time and thus represent a true family unit. The characteristic Hertzsprung–Russell diagram then emerges from the different rates at which the various stars evolve and the age of the cluster can be ascertained from the position of the left-hand cut-off point from the main sequence. The youngest clusters, like h and χ Persei are only several million years of age whereas NGC 188 is around one billion years old. As no older open star clusters have been found, there must be forces which are capable of breaking up the unit and scattering the individual stars throughout space. In fact, we do know of three processes which could do just that. Large interstellar clouds which either penetrate the cluster or surround it can change the force of gravity such that several stars become 'absent without leave'. Then there are stellar encounters between clusters leading to an exchange of energy, as a result of which kinetic energy can be transferred from one star to another. The 'winner' of this exchange can then move so fast that it overcomes the force of gravity and escapes from the cluster. As this process is similar to the physical changes which occur on liquid surfaces during vaporization, we say that the star is 'vaporized'. (Some atoms and molecules on the surface of water acquire such

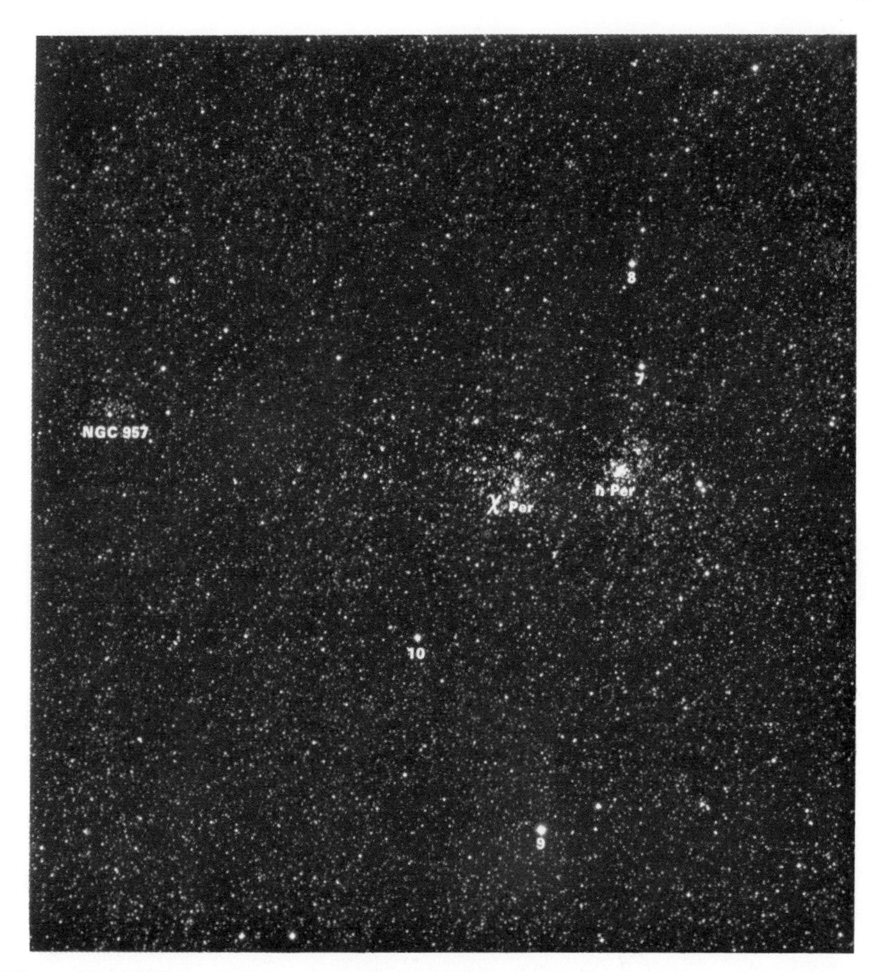

Figure V.10. The two open star clusters h and χ Persei in the Perseus constellation contain more than 300 stars up to stellar magnitude 15.5. These clusters are about 3 million years old and this puts them amongst the youngest objects in our star system. Both clusters are about 2250 pc away. (Photograph: H. Vehrenberg)

high velocities as a result of collision processes that they leave the liquid compound and thus vaporize.) The result of all this 'vaporization' in space leaves the star cluster smaller and more concentrated as the star number decreases. The third disruption process is the so-called differential rotation. The galactic clusters move in large orbits around the Milky Way centre, in the course of which those clusters nearest the centre have greater velocities than those further away and if a cluster is small and not compact enough, i.e. has insufficient self-gravitation, then it will probably be 'ripped apart'. Differential rotation is probably the most common cause of destruction amongst galactic clusters. In any event, the disruption mechanisms mean that over a period of time, stars broken away from their clusters will simply be lost in the general 'sea of stars' and no longer be recognizable as former members of a cluster.

Figure V.11. Open star cluster Praesepe (Beehive) in the Cancer constellation. Praesepe is one of the best known open star clusters and its brightest star, ε Cancri, has an apparent brightness of magnitude 6.3. It contains about 500 stars in total and is about 150 pc away. (Photograph: H. Vehrenberg)

A special form of star concentration was discovered in 1949 by the Armenian astrophysicist V. A. Ambarzumjan. He found so-called stellar associations or aggregates in some parts of the sky. The diameter of each association is about 100 pc and the density is not much greater than in its general surroundings, but whereas stars of various spectral types are all mixed up together in other areas, the associations are composed almost exclusively of early spectral types. Hence, concentrations of these spectral types, and O and B stars in particular, are extremely high and a very young open star cluster is often found right in the middle of an association. Radio observations have shown that the system occasionally becomes embedded in a large cloud of neutral hydrogen. Star associations are especially young objects; they do not date back further than a few million years and stars are even now being born in one of the most well

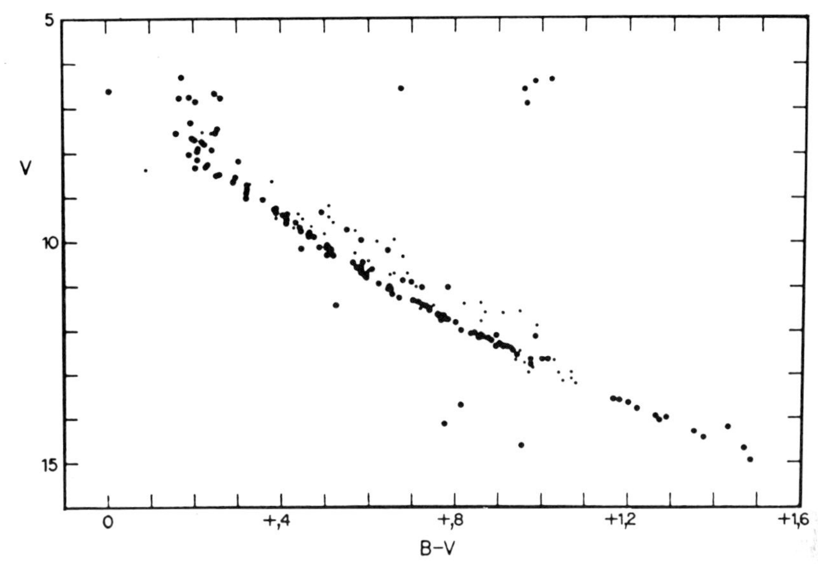

Figure V.12. Colour can be plotted on the Hertzsprung–Russell diagram instead of spectral type. Astrophysics uses various accurately defind colour systems—the difference in brightness in the blue and yellow spectral range is frequently used (B–V). This adapted Hertzsprung–Russell diagram is known as the Colour–Magnitude diagram. When plotted for the stars in the open star cluster Praesepe, the main sequence stands out clearly, with the break-off point at B–V = 0.2 and five stars already in the Red Giant area. Stars slightly above the main sequence line are probably binaries. (Diagram after H. L. Johnson)

known of all, in the Orion constellation. However, they are so loosely structured that they will all drown in the 'star sea' after a realtively short time and will cease to be recognized as members of a family unit. In some instances, this divergence can be recognized from the proper motion of the stars, and it is interesting to note that many stars are 'torn out' of the association and move away at great speed (in this case, they are known as 'stray' stars).

Globular Clusters

A second type of star family is represented by the spherical star cluster, or globular cluster for short. Relatively well known examples of this are the M3 clusters in the Hercules constellation and ω Centauri in the Centaurus constellation. Even outwardly they are very different form the open star clusters. Whereas the number of stars in the latter is modest, individual stars are easily recognized and the whole unit relatively loosely structured, even a cursory glance at the globular cluster will reveal that it is a regular, almost perfect sphere-shaped configuration with an incredibly high star content. Star counts along the edge of globular clusters—the only area in which individual stars can actually be seen—and corresponding extrapolation, give figures in the region of 50,000 stars for the smaller ones and up to 50 million for the larger varieties. Taking a diameter of 70 pc as a basis for the smaller 'star systems', the

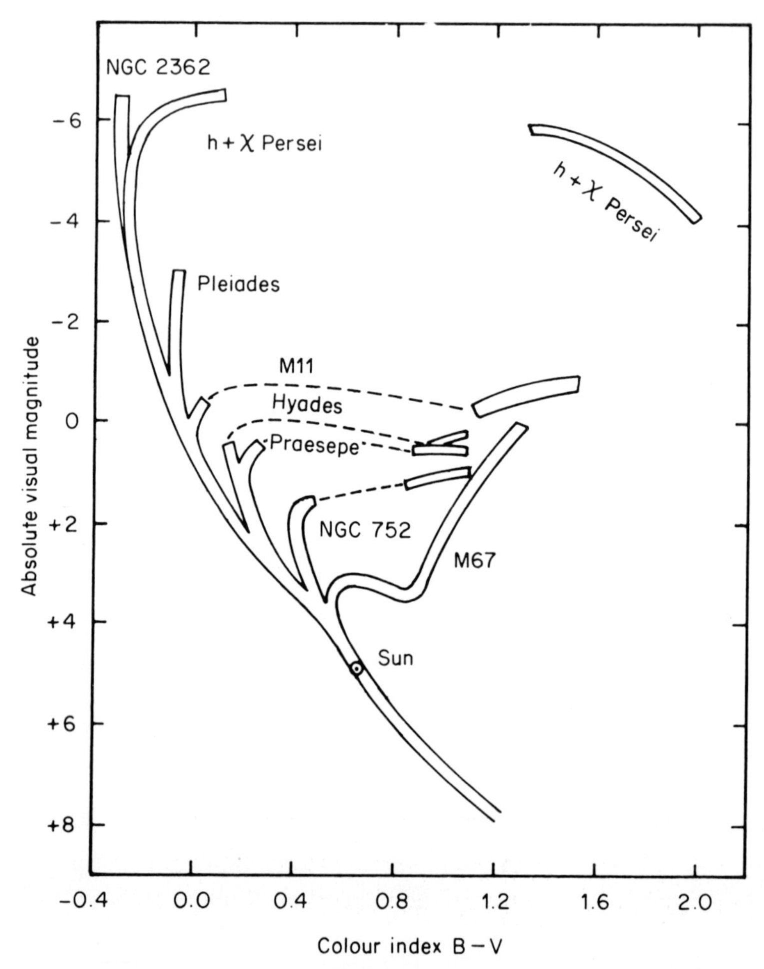

Figure V.13. When Colour–Magnitude diagrams of various open star clusters are brought together, the different break-off points from the main sequence emerge very clearly. This provides evidence that open star clusters do have different ages. h and χ Persei and NGC 2362 are especially young, whereas NGC 752 and M 67 are old (After A. Sandage)

star density seems to be 1000–10,000 times greater than that in the vicinity of the Sun. The density in the centre of a globular cluster is calculated to be approximately 1000 stars/pc^3.

Although the correlation between stars in a globular cluster is apparent even from the outside, the Hertzsprung–Russell diagram shows the true relationship. Once again, there is a main sequence, but it is a very short one and veers towards the top right, towards a branch of the Red Giant configuration. Just as in the case of the open star cluster, this type of Hertzsprung–Russell diagram represents an evolution picture; however, the open star clusters contained relatively young objects, whereas the globular clusters are made up of much

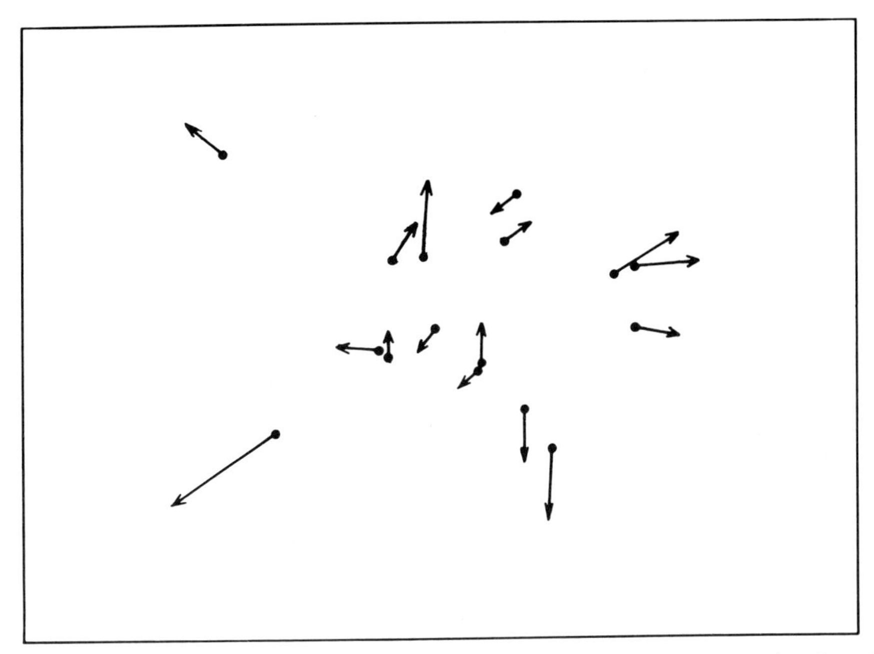

Figure V.14. The movement of stars in the ζ Persei association was measured and used to calculate changes over the next 500,000 years; the different locations were then plotted. It is clear that the association will either expand or split up

Figure V.15. Photograph of globular cluster M 3 (NGC 5272) in the Canis Venatici constellation. (Photograph: Mt. Palomar Observatory)

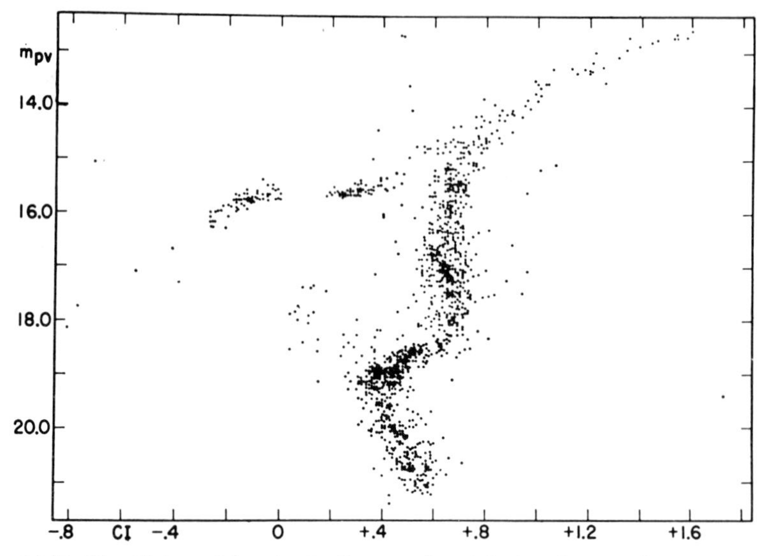

Figure V.16. The Colour–Magnitude diagram for globular clusters shows a very short main sequence. On the other hand, there is a steady movement towards the Red Giant area and further on towards the White Dwarf region. The course is broken only at B–V ≈ 0.1 where the RR-Lyrae type variables are found. The break-off point way over to the right and the high number of stars in the Red Giant area shows the extreme age of the globular clusters. (Diagram after A. Sandage)

older ones. Hence the Hertzsprung–Russell diagram does not show the upper part of the main sequence as all the massive stars have used up all their 'fuel' long ago and become Red Giants or White Dwarfs. Moreover, the clusters are so large and compact that disruption processes are not much in evidence, all the more as the cluster does not rotate around the centre on the same scale as the open star clusters and does not by chance become surrounded by interstellar clouds. Galactic clusters are about 8–12 billion years old and represent one of the oldest phenomena in the star system as a whole.

Star Populations

Stars differ from each other not only in their spectra; moreover, spectral classification does not cover such important facets as star family membership, heavy element content as part of their chemical composition (the so-called metal abundance) and properties of motion. The last aspect is in itself a decisive factor in classifying the various star populations. Although members of these star races are often mixed up with others in space, each community forms its own particular system based on special properties of motion (covered in a later chapter), chemical composition and age. The most important star population groups are as follows:

1. The spiral arm population I, characterized by very young, hot, highly luminous stars with a relatively high metal content in their chemical

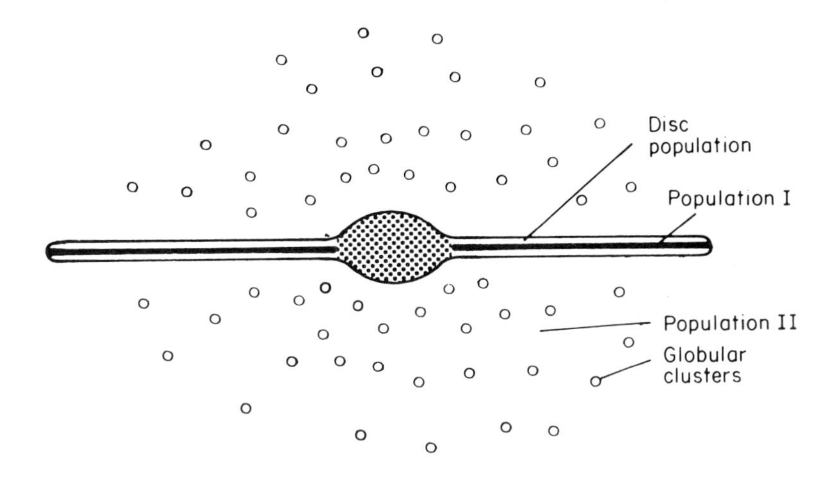

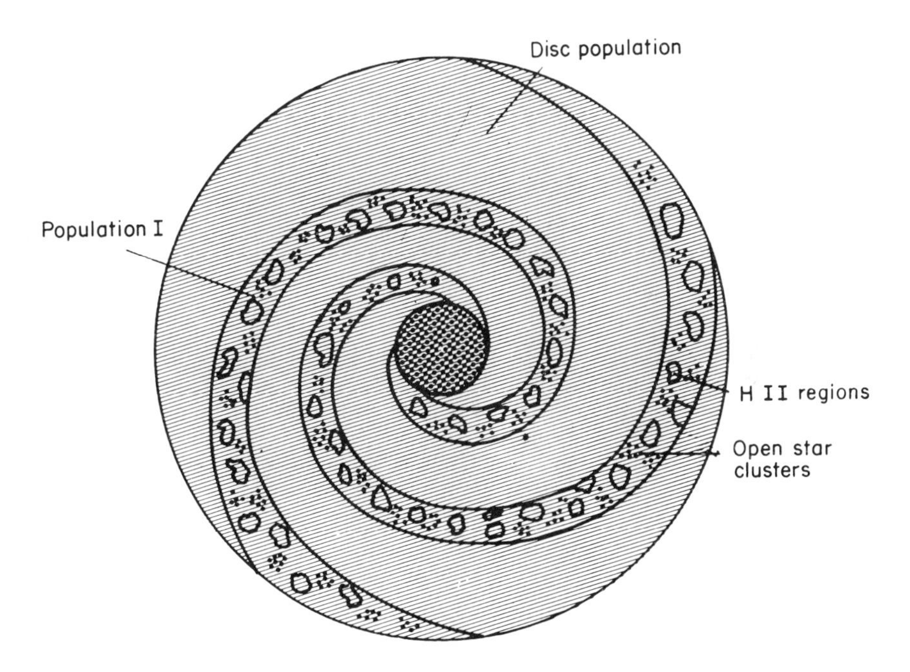

Figure V.17. The various populations are distributed around the star system in a very characteristic way. Spiral arm population I is densely clustered around the galactic plane (side view) but occurs only in the spiral arms themselves. On the other hand, the disc population is scattered over the whole disc and is not so highly concentrated around the plane. Halo population II is almost non-existent around the plane but tends to accumulate around the centre

make-up. A good deal of energy is radiated in the blue spectrum, hence they are termed 'blue stars'. Young open star clusters and associations and interstellar matter belong to this group. In large star systems, members of the spiral arm population I are arranged in a spiral arm formation.

2. Most stars in the general star field, especially those in the Milky Way, belong to the disc population. The members are arranged in the Milky Way disc and move in circular orbits around the centre. Our Sun seems to be a typical representative of this star group.
3. Hal population II is made up of globular clusters, RR-Lyrae stars and the so-called metal-impoverished subdwarfs. A particular property of this group is their low content of metals which are heavier than helium; metal abundance is only one five hundredth to one fifth of that of the normal figure for stars in the disc or spiral arm populations. Population II stars belong to a large, not very flat system—the halo—in which members are concentrated in the centre with a long orbit stretched around. Members of halo population II are about 10 billion years old.

Now that we have isolated the stellar population into individual and binary stars, family members or part of a whole race or population, we can turn our attention to general distribution patterns, the dynamic properties of celestial objects and the actual structure of the heavens.

VI

Order or chaos?

The Galactic Coordinate System

Any investigation into the structure of the heavens is closely associated with the question: which objects are to be found where? Do the various 'inhabitants' have specific places of origin or are they scattered and intermixed in the Universe at random? The celestial coordinates and distance of each object must be found before any accurate assessment can be made. Unfortunately, distance calculations are often laborious and not very accurate. However, as Herschel's star count showed, celestial coordinates give a good deal of valuable information. Hence the first step is to incorporate the objects, e.g. dark nebulae, open star clusters or galactic clusters, into a coordinate system and to estimate concentration. The galactic coordinate system has been found to be the most satisfactory. The network is similar to the grid map on Earth, except that it is projected into the celestial sphere and the equator and poles are in a different position. The coordinates right ascension (α) and declination (δ) frequently used in astronomy correspond to the earthly grid map projected into the sphere, in which the correlation geographic length = right ascension and geographic width = declination applies. However, galactic coordinates can be found by placing the equator plane of the coordinate system across the great circle of the Milky Way. Counting on the galactic equator begins in the Sagittarius constellation, the north pole is in the Coma Berenices and the south pole in the Sculptor constellation. Just as the Earth's coordinate system is plotted on paper in the form of an atlas, the galactic coordinate network can also be plotted on a flat surface to give the characteristic pictures reminiscent of the grid lines on world maps. The horizontal symmetric lines represent the galactic equator and correspond approximately to the Milky Way belt in the heavens. All points around are in either the galactic or the Milky Way plane. If a celestial equator, i.e. a projection of the Earth's equator, were plotted in this coordinate system, the line would emerge in the form of a wave.

Distribution of Objects in the Galaxy

Star clusters and associations are the most conspicuous objects in the heavens and therefore their distribution will be considered first. A catalogue published

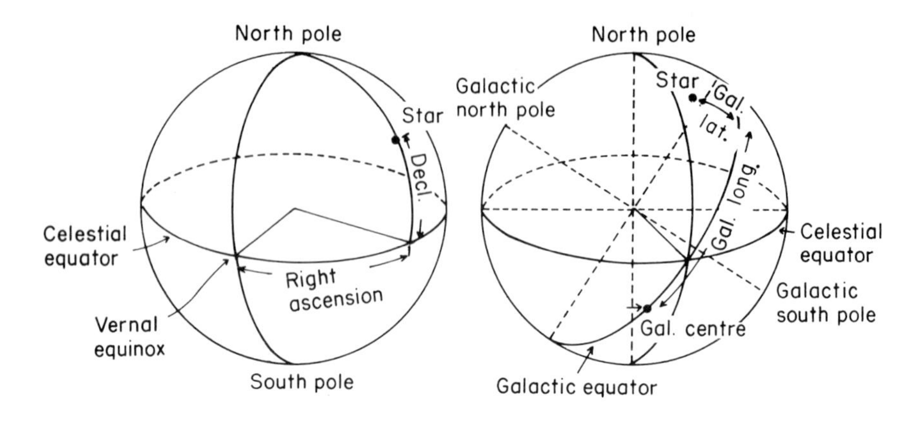

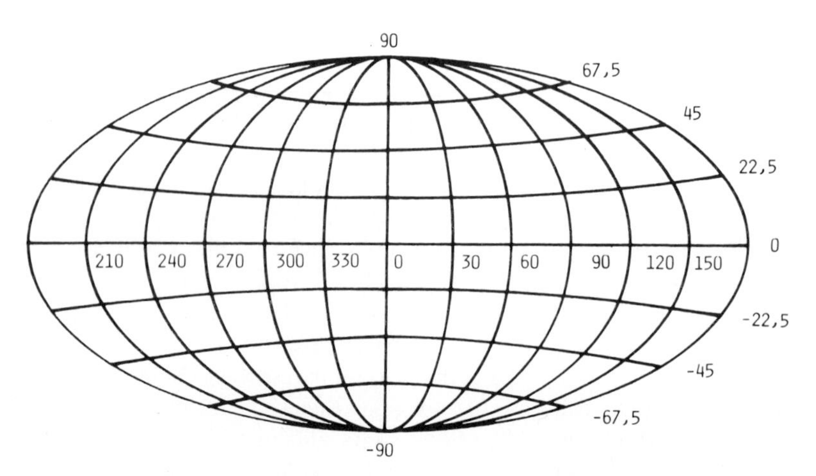

Figure VI.1. A coordinate system frequently used in astronomy is the equator system, the datum plane of which is the extended plane of the Earth's equator. The longitudinal axis is the polar axis (α) and the latitudinal one is the declination axis (δ). In this system, the Milky Way is inclined towards the equator as shown on the right. If the coordinates of a star are plotted on this datum plane, these are then the galactic coordinates (l = length, b = breadth). The starting point for calculations on the galactic equator is the galactic centre (seen from the Sun). In the equator system, this starting point is at the coordinates α = 17 h 42 m 37 s and $\delta = -28°57'$ (1950). When this galactic coordinate system is put on paper, it forms the 'galactic world map'

in 1970 listed a total of 1055 open star clusters, 70 associations and 125 galactic clusters. This is certainly by no means an exhaustive list but represents all those discovered so far, viewed from the Sun. Interstellar dust prevents us seeing into the 'profundity' of space, but it is generally assumed that there is a total of 10,000 to 100,000 open star clusters and up to 10,000 associations. All associations and open clusters are in the immediate vicinity of the galactic equator and are, therefore, inhabitants of the galactic plane. The concentra-

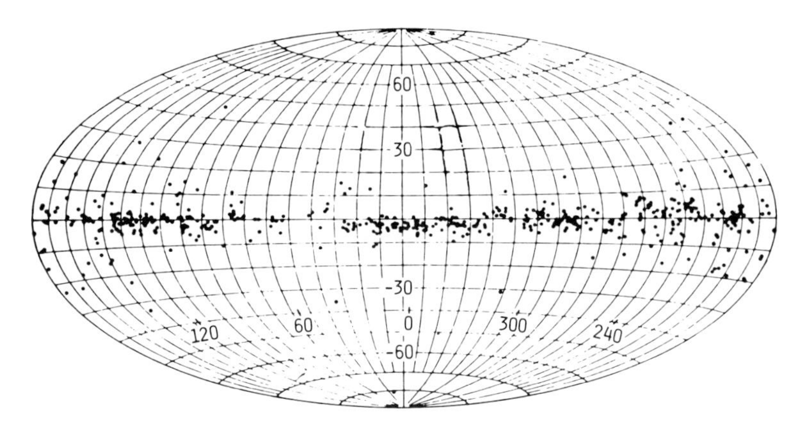

Figure VI.2. Distribution of open star clusters in a galactic coordinate system (according to S. Sharpless). In this distribution, concentration falls out at the galactic centre and at $l = 125°$ and $l = 210°$, where the inside of a spiral arm is visible. The gaps between $l = 30°$ and $l = 50°$ are caused by an intervening dust complex

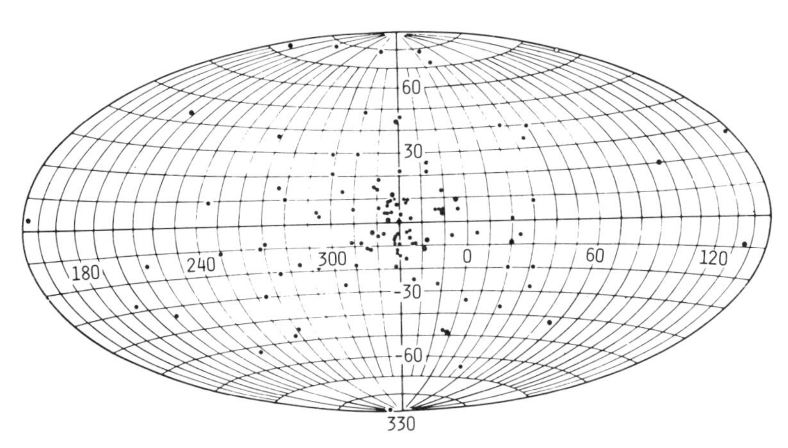

Figure VI.3. Galactic distribution of globular clusters (according to Hogg). Shapley concluded that in view of the concentration of globular clusters around the galactic centre and the relative vacuum in the anticentre ($l = 180°$), the Sun could not be in the centre of the system. The globular clusters are not restricted to the galactic equator zone

tion between 290° and 30° galactic longitude and the significant gap between 30° and 50° are two striking factors. Several clusters in the mean galactic latitude are found between 100° and 230°. Further investigation into these clusters showed that they are not far away from us and thus deviation from the galactic plane is only a simulated effect.

On the other hand, galactic clusters have a very different pattern of distribution. They are much more uniformly spread out and many are found in high galactic latitudes. There is no pronounced concentration on the galactic plane, but a specific build-up around the coordinate's origin, the galactic centre. In view of the great distance involved, this is certainly no simulated effect—galactic clusters are more uniformly distributed in space than open star clusters.

The H II regions—and there are probably in excess of 10,000 of these—show a similar distribution pattern to open star clusters. Again, the concentration falls almost exclusively along the galactic equator. However, density is greatest around the galactic centre with gaps to left and right. The left gap is already familiar from open star cluster observations and is apparently real, but the right gap is virtually in the exact position of the dark cloud known as the 'Southern Coalsack'. This becomes more apparent when the dark clouds listed in a large catalogue are plotted in a similar way. Again, the most salient characteristic is the high concentration along the galactic plane. However, there are no gaps between 30° and 50° galactic longitude and the dark clouds in the 140–220° galactic longitude range below to the Sun's environment (for example, the Taurus complex, longitude 165°, latitude −16° is no further than 250 pc away).

The distribution of interstellar matter has been assessed by observing dark and bright cloud patterns and can be recorded in the radio range. For example, the distribution of neutral hydrogen has been derived from study of the 21 cm line. Its distribution pattern is essentially similar to that of other interstellar matter. Continuous radio frequency radiation also shows a concentration around the galactic plane but again, distribution around the galactic plane is not fully uniform. Although the symmetric point of the particularly intensive central region appears around 0° galactic longitude, there are deviations, particularly at $l = 30°$, and minimum intensity is around 240°, not 180° as expected. Apart from these disc components, as radioastronomers call them, uniform radiation comes from the whole celestial sphere. However, radio telescopes with particularly sensitive directional response have shown that most of this radiation comes from sources outside our star system (so-called extragalactic radio sources).

The dust contents of interstellar matter not concentrated in the dark clouds has a similar distribution pattern to the H II regions and the dark clouds. This emerged from a nebula count carried out by the American astronomer E. Hubble. These nebulae are extragalactic star systems totally unrelated to the normal stars of H II regions and have therefore been passed over when considering the celestial population. They are foreign 'star heavens' or 'cosmic islands'. Hubble investigated these objects and found that they are distributed relatively uniformly throughout space, except for the Milky Way where they are barely in evidence at all. The only explanation for this original concept is that the light-absorbent dust in our star system prevents us from looking 'outwards' into this layer. Thus, finely distributed dust grains are concentrated in the galactic disc.

The various distribution patterns show that most stars and interstellar matter are arranged in a relatively thin disc formation. However, the exact structure of the system remains a mystery until the distance from the various celestial objects is brought into the equation. In many cases, this means that some knowledge of star movement must act as a basis.

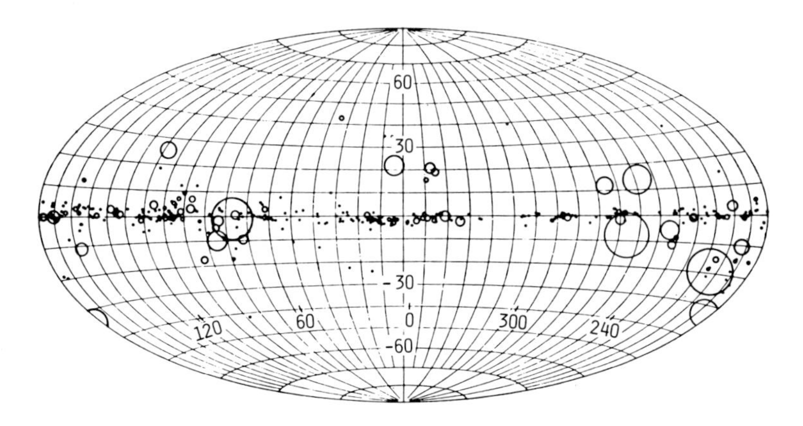

Figure VI.4. Distribution of H II regions in a galactic coordinate system (according to S. Sharpless). The distribution pattern of H II regions is a mixture of concentrated areas and empty spaces. The southern Coalsack or Scorpius–Ophiuchi complex is between l = 300° and l = 330°. The concentration at l = 290° is joined with the Sagittarius arm and that at l = 80° or l = 27° with the Perseus arm, whereas the gaps between l = 25° and l = 60° represent inter-arm areas. The circle size is a rough measure of the angular diameter of the H II regions seen

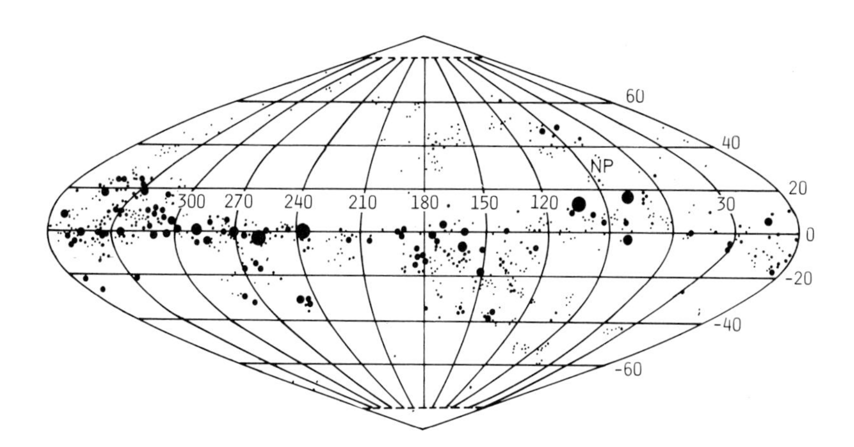

Figure VI.5. Distribution of dark clouds in a galactic coordinate system (according to Lundmark). This drawing differs from the others in that old galactic coordinates are used. So the galactic centre is on the left at l' = 330° and there is a great dark cloud complex to be seen there. A second one, the Scorpius–Ophiuchi complex, is at l' = 335° and b' = 20°. The Orion-Taurus complex is at l' = 150° and b' = −20°. The dark clouds also tend to concentrate around the galactic plane. As the Orion–Taurus complex is not very far away from the Sun, the deviation from the Milky Way plane as seen here is misleading

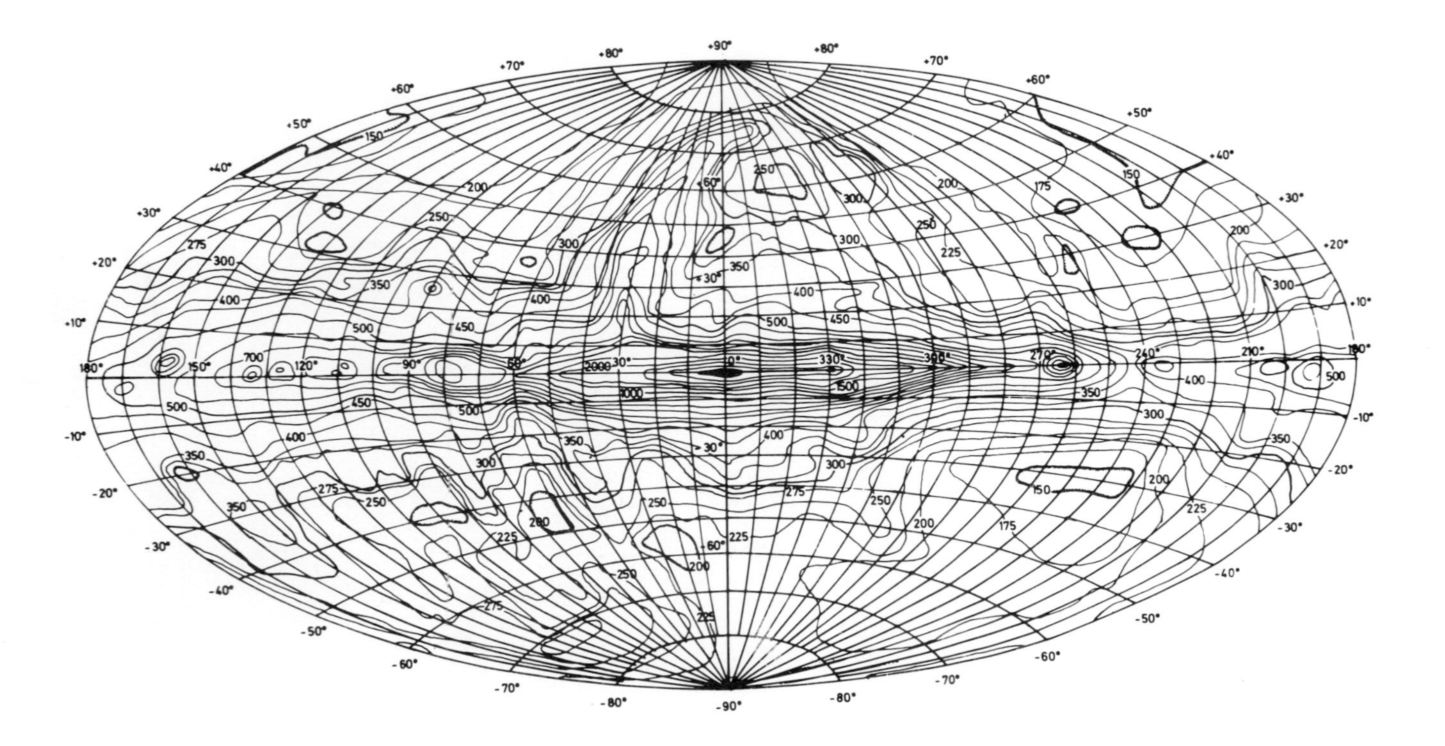

Figure VI.6. Radio map at 150 MHz in a galactic coordinate system (according to Landecker and Wielebinski, 1970). This map shows lines of similar intensity (isophots) as formed by 150 MHz radio frequency waves; the numbers are a measure of intensity. The most important characteristics of this map are the concentration in the centre, the radiation maxima at l = 80°, 130°, 210° and 265° and the general concentration of radiation along the galactic plane. At 80° and 265° we can see into the Perseus arm and at 210° the angle of vision runs tangential to the Orion arm. The trail northwards at 30° is probably a local effect, possibly caused by remains of a Supernova explosion

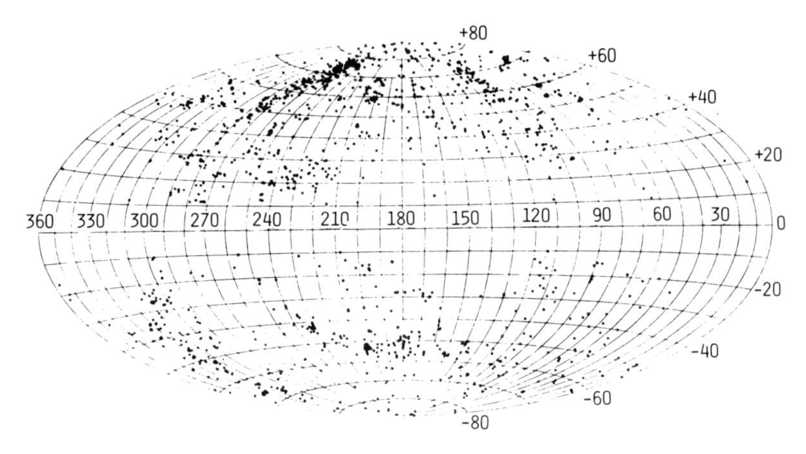

Figure VI.7. This figure depicts the 1249 galaxies listed in the Shapley–Ames catalogue as a galactic coordinate system. No galaxies are seen in the equator zone between $b = -10°$ and $b = 10°$, but otherwise they are reasonably uniformly distributed. This galaxy-free zone is due to the light-absorbent dust layer in the galactic plane. (Harvard Observatory diagram)

The 'Unchanging' Heavens are Full of Movement

A study of the heavens over several years will show very little change. The planets with their changing positions, comets and meteors are of no further interest to us as members of the planetary system, as far as this survey is concerned. The variable stars and the Novae, which are very seldom seen (a Nova did appear in September, 1975, in the Cygnus constellation), are the most significant. However, the sky is not totally fixed, and the positions of the stars in relation to each other do change continuously over a long period of time. This visible stellar motion in the celestial sphere, called the *proper motion* of stars, covers only movement *vertical to the optical path.* In reality, however, the stars 'criss-cross' through space, so a second component of motion, which lies *in the optical path,* must be found before spatial stellar motion can be determined. The Doppler effect can be used to measure the radial velocity and this, together with proper motion and distance, forms the basis on which spatial velocity can be calculated.

The proper motion of very many stars is known and is based 'solely' on photographs taken on the heavens on two specific occasions with as much time in between as possible, and then a comparison is made of the star positions in each. However, proper motion is an extremely small unit and often involves a degree of uncertainty, so that the 300,000 and more known values for proper motion are a basis for statistical calculations. Only about 100 stars move by more than 0.1 arc second per year (0.1 arc second of proper motion per year means that the star takes 18,000 years to pass a full moon width across the sky).

However, the star spectrum is required for radial velocity calculations, as spectral line displacement has to be compared with laboratory wavelengths.

These tests are very time consuming and as a result the radial velocity of 'only' 20,000 stars is known to date. Most radial velocities are about 20 km/s and the upper and lower limits are +543 and −389 km/s, respectively (all motion is relative to the Sun). Apart from some slightly 'muddled' star motion figures, the following interesting facts have emerged: stars around the point at coordinates $\alpha = 270°$ and $\delta = 30°$ (Hercules constellation) have the greatest negative radial velocity and have no proper motion in general. Stars directly opposite have the highest positive radial velocity, but again, proper motion is insignificant. The stars vertical to the line of contact between these two points have the greatest proper motion and no radial velocity. Thus, it seems as though all stars leading up to the Hercules constellation are coming towards us, those leading from it are running away and those vertical to the contact line are simply 'scurrying by'. If the concept of a static Sun is abandoned, then these findings can be explained simply by the movement of the Sun towards the Hercules constellation at a speed of 20 km/s. The goal is called the apex and is located at coordinates $\alpha = 270°$ and $\delta = +30°$ and the movement itself is called the peculiar motion of the sun.

The Star Carousel

Other stars have a peculiar motion too, but it seems to be a relatively random and irregular occurrence and is, therefore, simply 'averaged out' for the larger star groups. This leaves just the systematic velocity rates to deal with as they cannot be fully explained by the movement of the Sun. In fact, if the Sun's motion is subtracted from that of the stars, then movements with highly original behaviour patterns remain. Stars in the $l = 0°$, 90°, 180° and 270° alignment have no components of motion in the optical path (radial velocity is zero), but stars in the galactic longitude of 45°, 135°, 225° and 315° have extreme radial velocities. When radial velocity values are plotted as a function of galactic longitude, a double wave emerges. Proper motion rates also give a double wave, but one which is at an angle of 45° to the radial velocity one and is positive only at maximum value levels.

There is a plausible explanation for this behaviour, but first let us describe an imaginary experiment. Journeys into space have confirmed that satellite movement obeys Kepler's Law and a satellite needs more time to orbit the Earth the further it is away from us. Thus, a satellite 200 km away takes 90 minutes, whereas one 36,000 km away takes a whole day. Imagine we are sitting in a spaceship which circumnavigates the Earth in 125 minutes and is 2000 km away. Four satellites will be visible from our ship, one a distance of 500 km and an orbit time of 95 minutes (situated 'under' the spaceship), one 'above' the spaceship, 3500 km away and with an orbit time of 160 minutes, and a satellite 'in front of' and 'behind' the spaceship. The last two satellites have the same orbit data as the ship itself. After only a short time, it will become obvious that the 'lower' satellite is gaining ground, the 'top' one is lagging and the 'back' and 'front' ones remain on the same level as our ship. Examination of

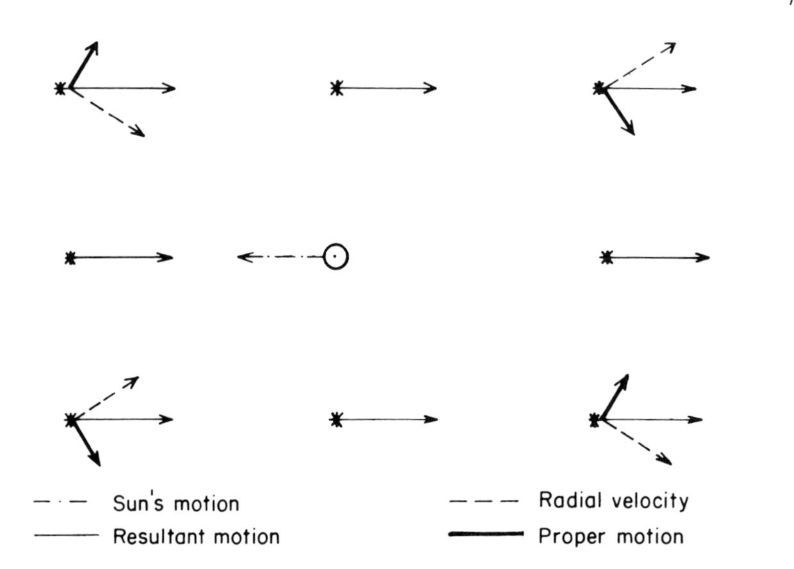

Figure VI.8. The movement of the Sun towards the Hercules constellation simulates movement of neighbouring stars in the sky. It seems as though the stars are diverging in this direction and coming closer together in the other. This apparent star movement becomes clear when the Sun's velocity is split up into its component parts, radial velocity and proper motion. The peculiar motion of the Sun is derived from observations of apparent motion: the Sun moves towards the point with coordinates $\alpha = 270°$ and $\delta = 30°$ at a rate of 20 km/s

radar echoes would show us further that, at the moment of overtaking, the radial velocity between satellite and spaceship is zero, but that proper motion is exceptionally high. However, radial velocity can be recorded both before and after overtaking. The radial velocity for the satellites 'in front of' and 'behind' the spaceship is zero, but path curvature does give rise to proper motion. Now, if both components of satellite motion are plotted as a function of optical path, the corresponding points will form a double wave and an explanation for the remarkable phenomenon of double wave motion has thus been found. The Sun corresponds to the spaceship, the Earth to the still unknown galactic centre, the 'upper' satellite to stars at $l = 180°$ (at the moment of overtaking), the 'lower' one to stars at $l = 0°$ and the satellites 'in front' and 'behind' to galactic longitudes $l = 90°$ and 270°, respectively. Just as satellite movement is affected by the pull of gravity from the Earth, so the gravitation from the galactic centre controls stellar motion; the further the stars are from the centre, the slower they are, and the closer they are to the centre, the faster their orbit becomes. In this way, the inner stars are always overtaking the outer ones and hence a double wave appears when looked at from the Sun, which is also revolving around the same centre. Stellar motion figures also help to define the Sun's motion more accurately. The Sun rotates round the galactic centre at a distance of about 10 kpc and it takes about 250 million years to perform one circle. It is, thus, travelling at a rate of 250 km/s.

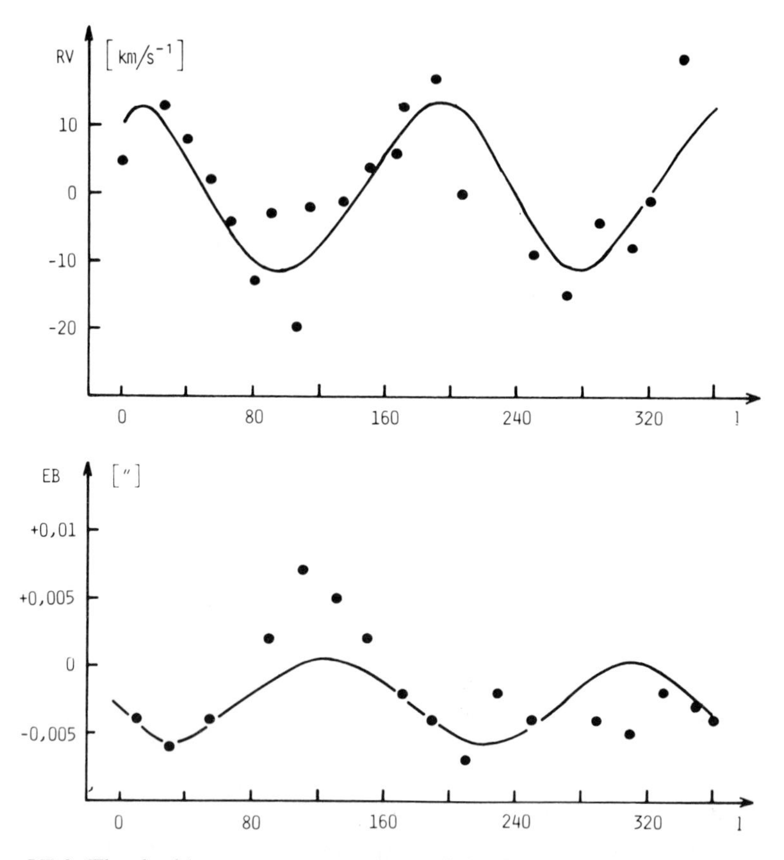

Figure VI.9. The double waves representing radial velocity and proper motion of the stars (corrected with respect to the Sun's motion) are proof that the star system, including our Sun, is involved in a differential rotation process. The proper motion double wave is phase-lagged with that of radial velocity by about 45°. As proper motion is difficult to observe accurately, the double wave here is not as pronounced as that of radial velocity. (After Plaskett, Pearce and Oort)

Interstellar dust grains prevent us from seeing all but a very small part of the galactic disc and so knowledge of rotation in this area must be applied to the whole of space. However, radioastronomy can be of enormous assistance in this field, with the 21 cm line of neutral hydrogen and recombination lines in H II regions being of special use. First of all, nothing is known about the distance of neutral hydrogen clouds or of the distant H II regions which are outside the optical field of vision. However, by assuming that they also take part in the galactic rotation process, radial velocities, calculated from Doppler displacement, can be used as a basis for distance estimates. The radial velocity double wave of the 21 cm line proves that this assumption is correct. Moreover, as the greatest and smallest radial velocities are recorded at the point at which the optical path forms a tangent with the orbital path, a graph can be drawn in which maximum radial velocity (or orbital velocity in this case) is plotted

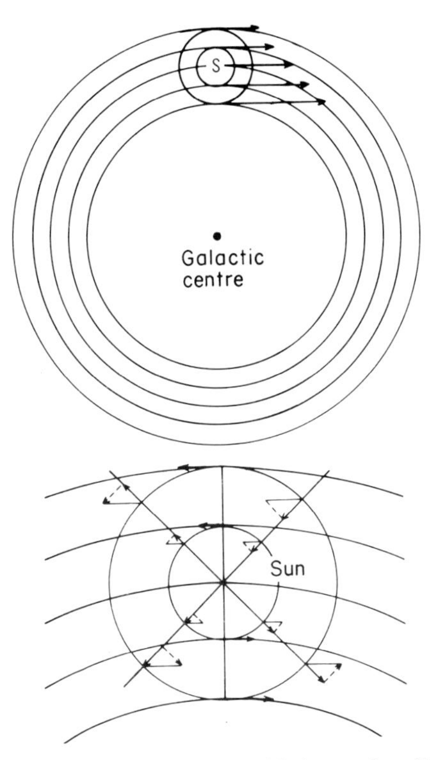

Figure VI.10. Rotation velocity decreases with increasing distance from the galactic centre, hence the radial velocity and proper motion distribution shown in the figure (double wave). This becomes very clear when star velocity is split up into its component parts (radial velocity and proper motion) for various sight lines

against distance from the centre. This galactic rotation curve is the key to determining the mass of the whole stellar system. Thus, study of stellar motion has yielded three important discoveries:

—the rotation velocity of the members of the star system is a function of their distance from the galactic centre (differential rotation);
—the distance of invisible interstellar matter can be estimated by using the laws of rotation in association with the 21 cm and recombination lines;
—the mass and mass distribution of the star system can be derived from the galactic rotation curve.

So far, we have mentioned stellar motion only in general terms; however, we know that each star differs from the next. How do star groups differ in their movement patterns and are there the usual exceptions to the rule? First of all, globular clusters seem to make very little contribution to the rotation process. Then there are certain groups, the RR-Lyrae stars in particular, which follow very eccentric orbits around the centre in a systematically different manner. The so-called high-velocity stars have a special role. These are stars not too distant from the Sun which have a velocity in excess of 65 km/s, relative to the

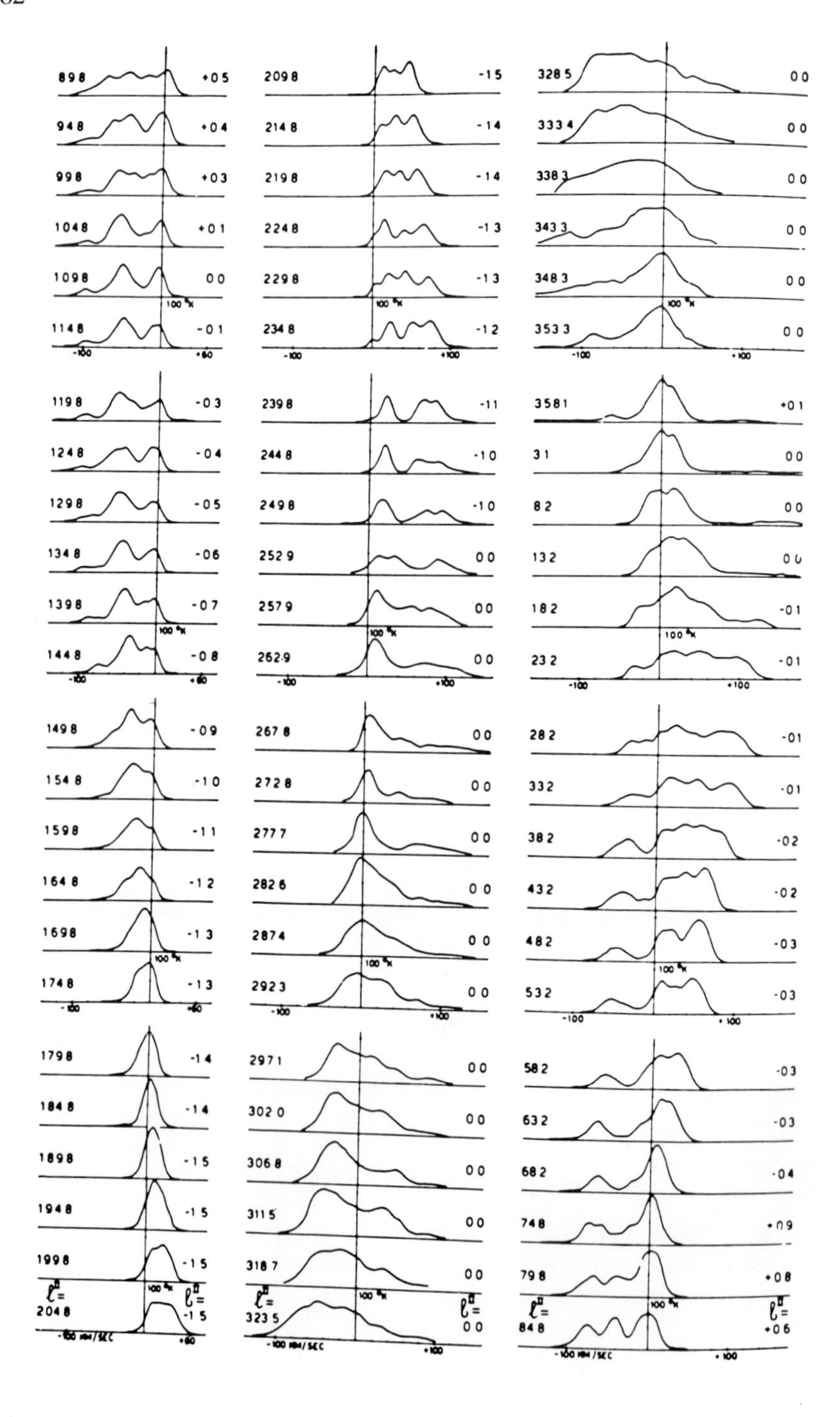

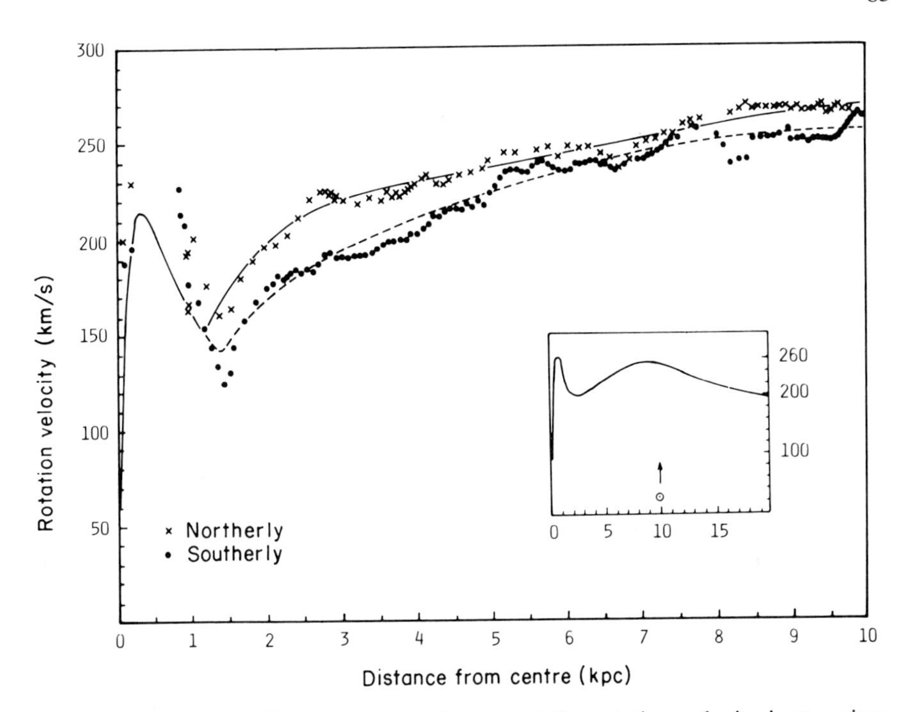

Figure VI.12. The rotation curve shows how great the rotation velocity is at a given point from the galactic centre. It follows from this that the innermost part ($R < 0.8$ kpc) rotates as a rigid disc (straight ascent), i.e. the rotation speed increases linearly with increasing distance from the centre. Then, up to about 1.5 kpc the velocity decreases with distance. This rapid reduction implies a break in continuity between rotation in the innermost central area and that in the nuclear disc. From 1.5 to 10 kpc the rotation velocity increases again, then drops once more, but gradually this time. Although the rotation speed increases with distance in the range 1.5–10 kpc, this does not mean that rotation is fixed—the increase is much too small for that and is also non-linear. The rotation curve within the Sun's sphere ($R = < 10$ kpc) is well known. In this case, the differences in velocity between the areas occur mostly to the north and to the south of the galactic centre and the rotation curve emerges in the form of a wave; this latter is a result of the spiral structure which characterizes the system. The North–South variation is due to a combination of spiral structure and large pockets of gas currents. (After Rohlfs and Quiroga)

Sun, and which generally move contrary to the direction of rotation. Thus, seen from the galactic centre, their rate of rotation is less than 185 km/s, so they could actually be called 'low-velocity stars'. The fact that no stars have been discovered with velocities in excess of 350 km/s points to the presence of a galactic escape velocity. Just as an Earth satellite travelling in excess of 11.2

Figure VI.11. Interstellar hydrogen also takes part in the differential rotation process, as 21 cm line measurements showed. At galactic longitudes 0°, 90°, 180° and 270° the line maximum had a radial velocity of zero and at 45°, 135°, 225° and 315° it reached extreme values. This conforms with the stellar radial velocity double wave (From Kerr and Westerhout)

km/s will leave the Earth's gravitational field for good, so a star going faster than 350 km/s would move out of the star system. The high-velocity stars move around the galactic centre in an elliptical orbit and approach the Sun right in the apocentre, the part of their orbit furthest away from the centre, where their orbital path speed is at its slowest, and thus relatively high compared with the Sun. These stars are members of population II. Stars from population I and the disc population all conform to galactic rotation regulations.

We would like to conclude this chapter on stellar motion by describing two additional facets. Star movement has always been seen as a disc activity. However, many stars are inclined to move around the line perpendicular to the disc plane, although the movement is much smaller and the repelling force of the disc is so great that the stars simply 'flounder' around the central position. However, apart from a few special star groups in halo population II, they remain there. The second point is concerned with star orbits in the plane. These are often compared to Earth satellite or planetary orbits, but the comparison is rather poor as stellar movement is not controlled by a small solid centre. The stars are much more likely to move in the complicated gravitational field common to all stars in the disc and galactic centre, whereas the central mass (Earth or Sun) controls the mechanics of the Earth satellites and the planets almost single-handed.

The Galactic Disc Pattern

The Sun belongs to the disc population, lies within the galactic disc composed of very many stars and, together with its colleagues, it takes part in the rotation process around the galactic centre. However, the position of the Sun makes it very difficult to study the galactic disc itself as we can see only a very few kpc inside before interstellar dust blocks the way and hides the centre completely. In spite of this, research into spatial distribution of selected objects has shown roughly what this centre looks like. It contains the bright, spectral type O and B stars, which can be seen as 'galactic beacons' from relatively great distances away even through the dust, the open star clusters and interstellar matter which is heard as a 'radio whisper'. Distance from the Sun or from the centre and galactic longitude are the decisive factors in drawing up a galactic plane distribution scheme.

The star associations are contained in the galactic disc over an area 5–6 kpc in diameter around the Sun, in three arc-shaped strips 1 kpc wide. The Sun itself is along the lower edge of the centre strip. Young open star clusters and H II regions give almost the same picture. However, the 'middle-aged' and elderly open star clusters do not form this type of strip pattern at all. The 21 cm neutral hydrogen measurements give an even better insight into the galactic disc structure, which also extends beyond the range of the Sun. They are undisturbed by interstellar dust grains and allow a glimpse not only of the galactic centre, but also of the other side of the disc. When plotted into a suitable co-ordinate system, the resultant picture shows a series of arc-shaped

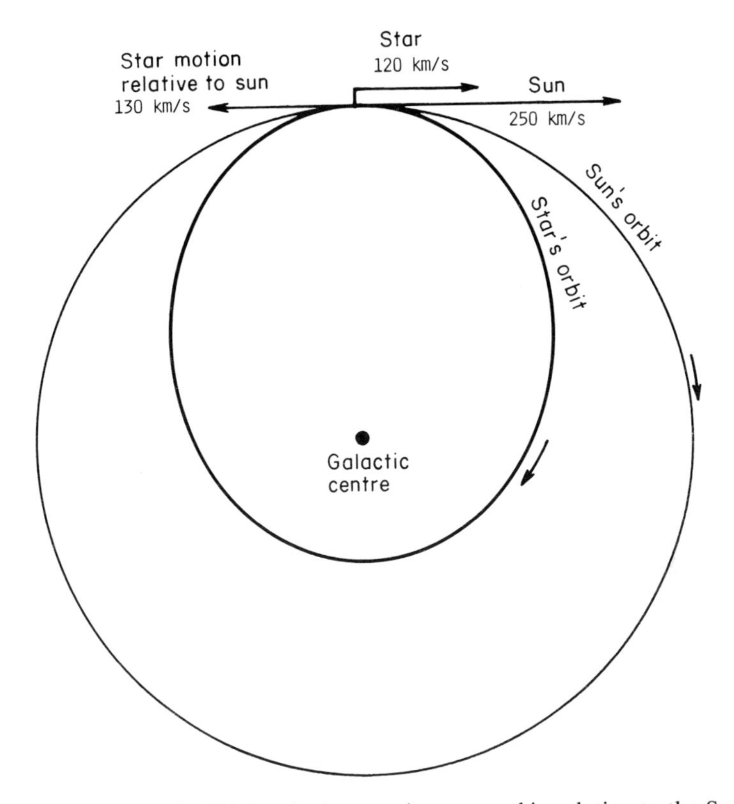

Figure VI.13. The speed of high-velocity stars is measured in relation to the Sun and is greater than 65 km/s. But these are in opposition to the direction of rotation, so that the 'genuine' speeds are less than 185 km/s. Thus 'low-velocity stars' are in elliptical orbit near the Sun but in the part furthest away from the centre, hence their low velocity figures

strips arranged around the galactic centre. More detail can be found in the vicinity of the Sun, as observation becomes easier. When comparing this distribution with that of associations, young open star clusters and H II regions, then it seems reasonable to assume that the strips form part of the curved hydrogen lines. When the picture is then compared further with other distant cosmic islands, the galactic disc pattern appears to be the spiral structure as we see it from our own 'part of the galactic carousel'.

In addition to the spirals, the centre forms an integral part of the galactic disc pattern. Although it is invisible to us in the optical range, it is nonetheless of extreme interest to radioastronomers. Infrared receivers have recently been introduced into this area of research. The radio review of the heavens already described had revealed the great radiation intensity from the central region in this spectral range. More accurate measurements soon showed that 'all kinds of things' are happening in the galactic centre. At about 3 kpc from the centre, there is a rapidly exanding ring of neutral hydrogen which is also rotating. Radioastronomers call this the '3 kpc arm'. On the opposite side of the 3 kpc

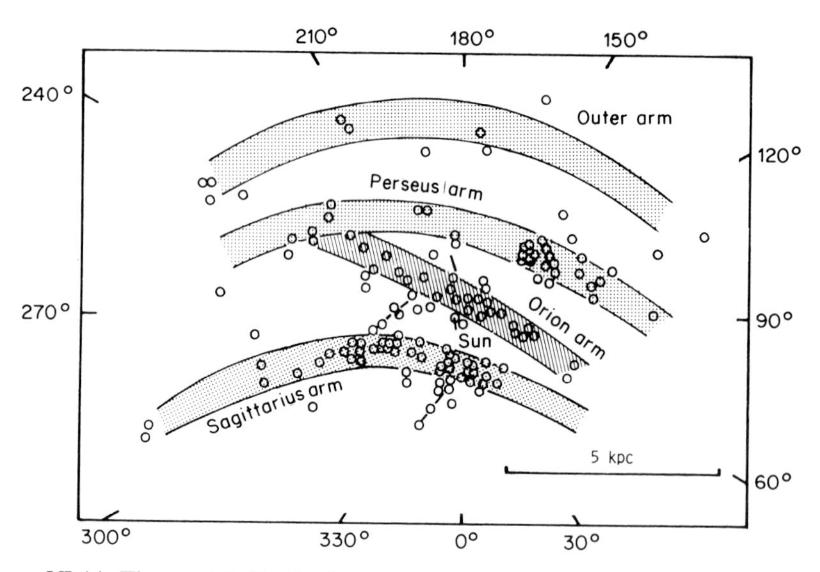

Figure VI.14. The spatial distribution of young star clusters and H II regions in the galactic plane can be seen as three 'strips', which represent parts of the spiral arms. The rather older star clusters have already 'wandered' out and are no longer part of the structure. Only very young objects are suitable spiral arm indicators. The spiral structure found in the vicinity of the Sun is in the form of three arms. The Perseus and Sagittarius arms are genuine spiral arms formed from a density wave, whereas the local or Orion arm, which contains the Sun along its inside edge, is not a density wave phenomenon. It is much more steeply inclined and probably does not belong to the spiral-shaped 'Grand Design' of our star system (After Vogt and Moffat)

arm (i.e. behind the centre from where we stand), the '21 cm ears' picked up the '135 km/s arm', which moves away from us at a rate of 135 km/s and also rotates. The 'nucleus-disc', also made of neutral hydrogen, is found between these two arms. It is obviously not a disc in the normal sense of the word, but a disc-like, rapidly revolving gas cloud, 1–1.5 kpc in size, enclosing another curved expanding gas cloud. Finally, there are a whole series of intensive radiation sources more or less in the centre itself. Some of these are thermal and some non-thermal in nature (a thermal radio source is generally an H II region and the synchrotron mechanism for the generation of radiation is the most predominant influence in a non-thermal source). The sources Sagittarius A and Sagittarius B are of particular importance. The name comes from the constellation in which the galactic centre is 'seen' from our position on Earth. The radio frequency source Sgr A is more or less dead centre and in fact consists of three radiation sources—the non-thermal east source and the thermal west source, which are both embedded in a larger thermal source. The structure is surrounded by a whole ring of molecule clouds including the largest of its kind, Sgr B2. This cloud ring is about 270–300 pc from the centre, expands at a rate of 100 km/s and rotates at 50 km/s. True 'molecule music' emanates from this part of the sky. Almost all molecules discovered in

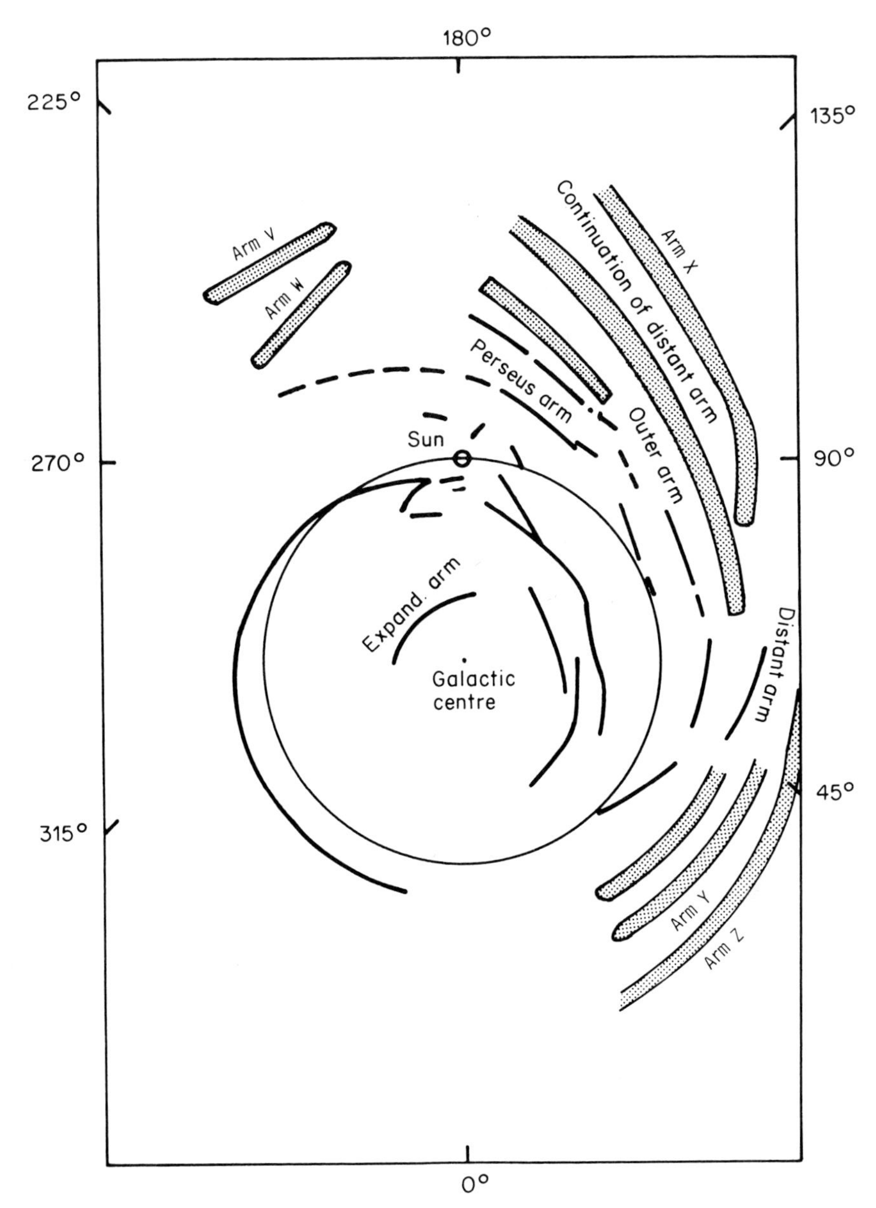

Figure VI.15. The most general methods used to investigate the spiral structure of the Milky Way system have their origins in radioastronomy. This figure is based on observation of neutral hydrogen and gives a clear picture of the long, stretched-out arc-shaped arms. The spirals are assumed to be part of our star system

interstellar space resound on the Sgr-B2 'molecule piano' which forms the largest known molecule cloud. It is not made up of exotic molecules, its main constituent being hydrogen molecules which are simultaneously embedded in H II areas with corresponding thermal radio frequency. What is more, Sgr B2

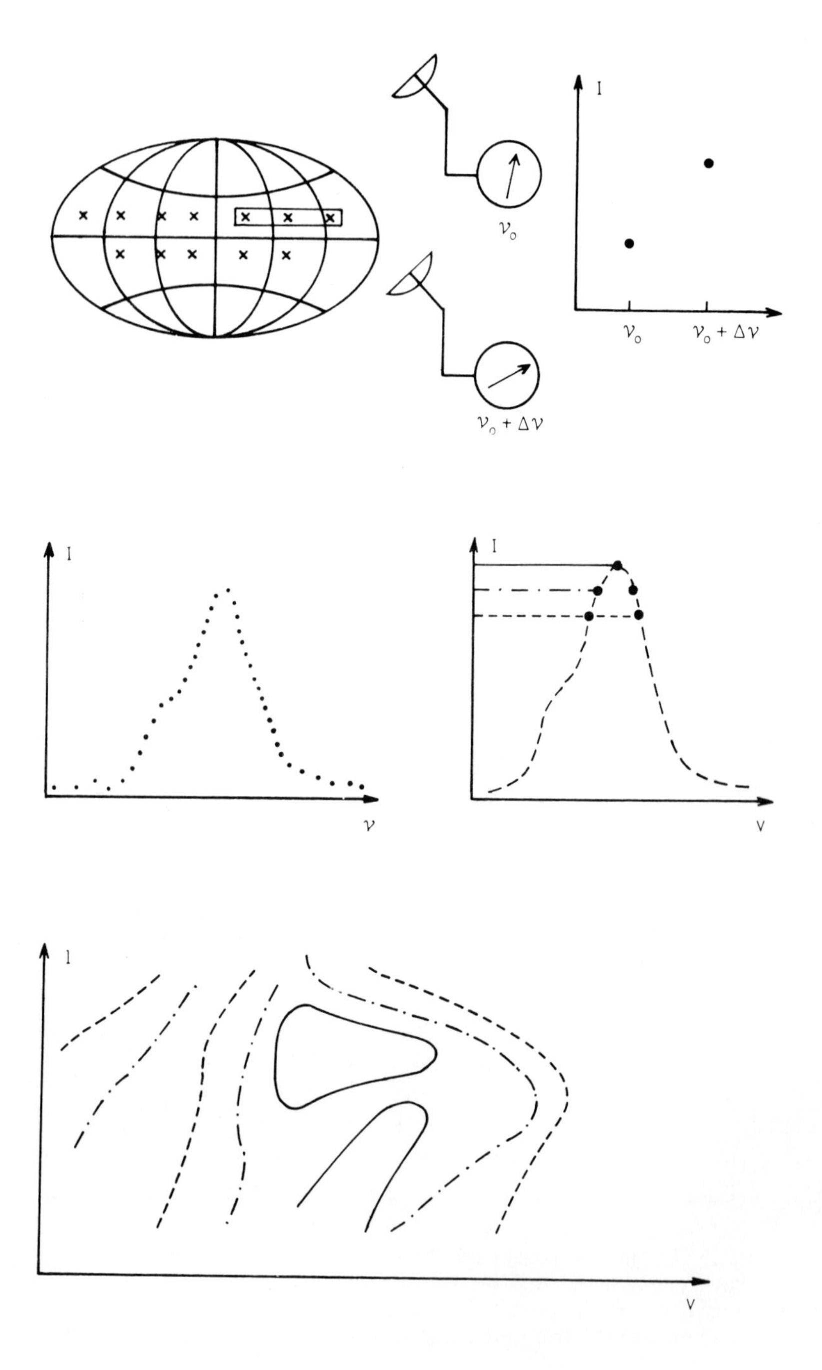
I
ν_0
ν_0
$\nu_0 + \Delta\nu$
$\nu_0 + \Delta\nu$
I
ν
I
v
I
v

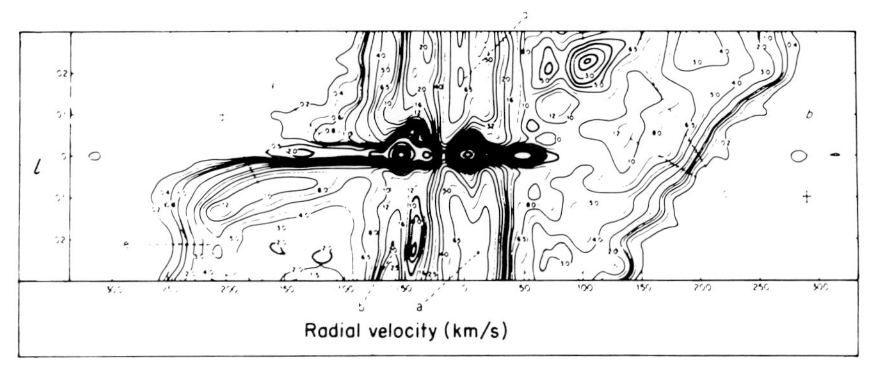

Figure VI.17. Longitudinal-velocity diagram of neutral hydrogen in the galactic centre. The great amount of hydrogen in the line of vision (a) is clearly shown, as is the 3 kpc arm (b), which also leaves behind its absorptive features in the radiation from Sagittarius B and which, therefore, must be positioned between us and the centre. As the arm velocity is negative, it is expanding and coming towards us. On the other hand, the 135 km/s arm is moving away (c). The elongated structures (d) and (e) indicate the rapidly rotating nuclear disc and (f) represents an expanded region in the nuclear disc. The hole in the middle is formed by the neutral hydrogen between us and the centre absorbing all suitable radiation from the Sagittarius A source (Diagram: Sanders and Wrixon)

and G 0.5–0.0 are positively giant H II regions and even these do not exist side by side in isolation but are enveloped in an expanded, thinner H II atmosphere.

Apart from the radio telescope with which we can 'eavesdrop' on the centre, there is also the infrared telescope which provides an 'eye' to penetrate the interstellar 'dust curtain'. Essentially three different infrared radiation sources can be picked up by it. The first is an expanded source measuring $3.6 \times 2°$, the second is $1.5 \times 0.5°$ in size and the third represents a series of discreet, smaller sources right in the centre. The first of these is apparently connected with the

---

Figure VI.16. The radio telescope shows us points of differing light intensity as opposed to the pictures which we see with our eyes. The word intensity is used rather than brightness as the latter is connected with visual observation. Thus, to observe the heavens in the light of the 21 cm line, we would proceed as follows: the radio telescope is adjusted to a certain point in the heavens and intensity is measured at frequency ν_0. It is then measured at a slightly different frequency, $\nu_0 + \Delta\nu$. This is pursued until the whole range of the line has been covered, then the radio telescope is moved to the next point and the process repeated. The individual measuring points from one point in the heavens combine to give the line profile. As we know that this line profile is due to the Doppler effect, the intensity at point ν_1, means that the gas complex is moving towards us, whereas at ν_2 it is moving away. At this point, we can regard it as velocity rather than frequency. A comprehensive picture of one section of the heavens is formed by plotting the same intensity for all absorbed lines as a function of their velocity and their actual position on a longitudinal-velocity diagram and joining all the points together. The lines formed are lines of the same intensity and often lines of similar hydrogen content also. These types of picture show how much hydrogen moves at a certain velocity in the line of vision and represent the key to understanding interstellar gas distribution in the Milky Way

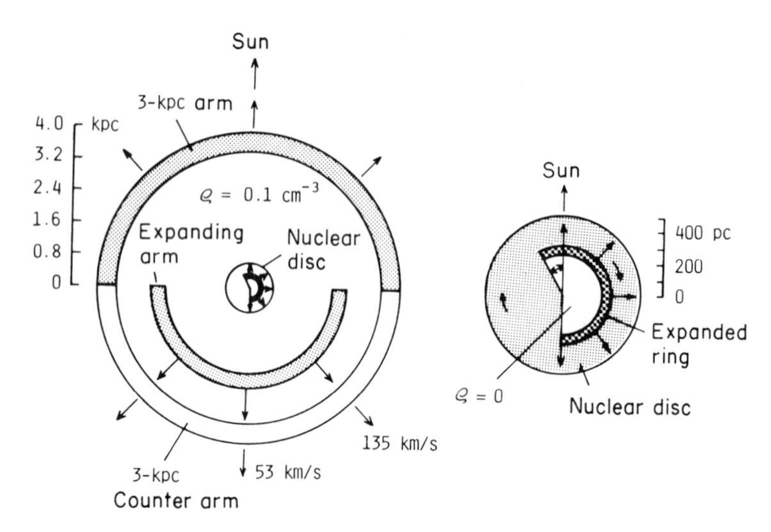

Figure VI.18. The model of the galactic centre, based on observations of interstellar gas in particular, consists of a fixed, rotating nuclear disc containing an ever-expanding ring. The disc is surrounded by the 135 km/s arm and the 3 kpc arm

great accumulation of stars in the centre. The many stars which are relatively tightly packed together in this area also emit a certain amount of infrared radiation. As the individual stars cannot be disentangled, the various small sources become 'blurred', resulting in the expanded infra-red region. The second source is connected with the expanded H II region already discovered in the radio range. The dust which is also present absorbs part of the stellar radiation, converts the shortwave radiation and re-transmits it in the infrared range. In this instance, the dust acts as a 'light transformer'. The smaller discrete infrared sources are connected with the H II regions *in* the expanded region and with the Sgr A and a non-thermal source (G 0.6–0.0).

Although we cannot study the galactic disc from 'above', various methods can be used to establish its pattern formations. Its main characteristic feature is the spiral-shaped structure with a high concentration in the centre. The spiral is emphasized throughout by bright H II regions, star associations, young, open star clusters and individual bright O and B stars.

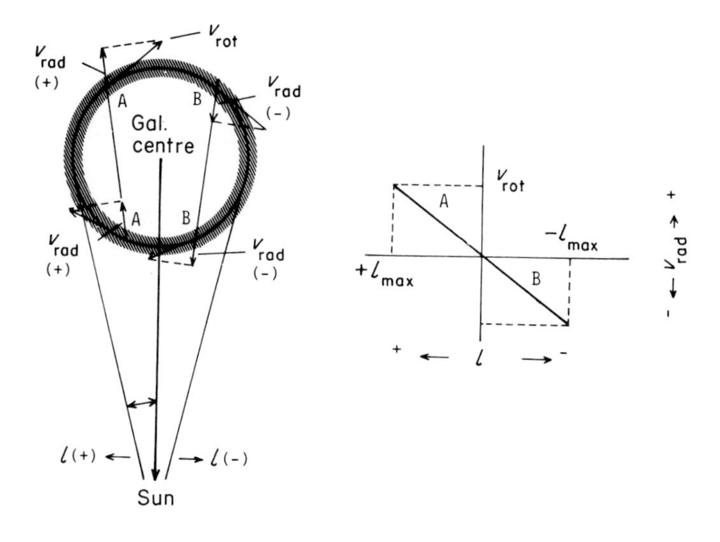

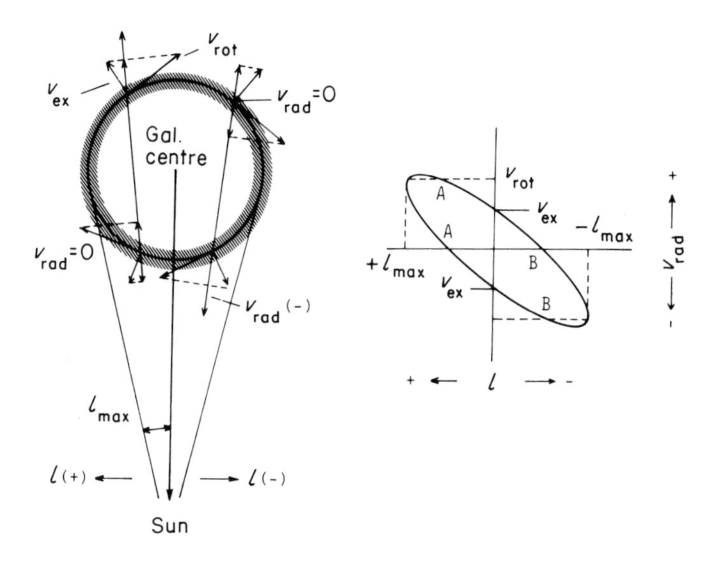

Figure VI.19. If the gas ring moves in a circle, diagonal lines of equal intensity emerge on the longitudinal-velocity diagram. If the ring not only rotates but expands as well, then the diagonal straight lines become elliptical or circular. The molecule cloud ring in the galactic centre emerged from this analysis

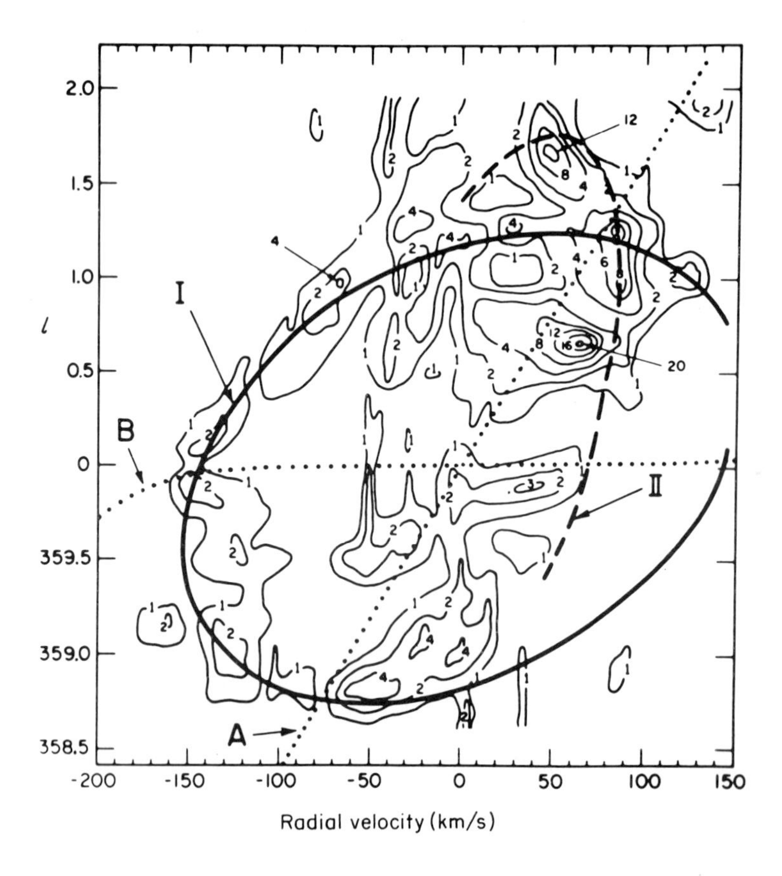

Figure VI.20. With some goodwill, the molecule line radiation maxima (the molecules in this case are formaldehyde, H_2CO) can be joined to form an ellipse in the longitudinal-velocity diagram. This ellipse represents the 'view' of the rotating and expanding molecule cloud ring in the galactic centre. The ring is relatively well safeguarded as similar structures have been found for other molecules. (After Scoville)

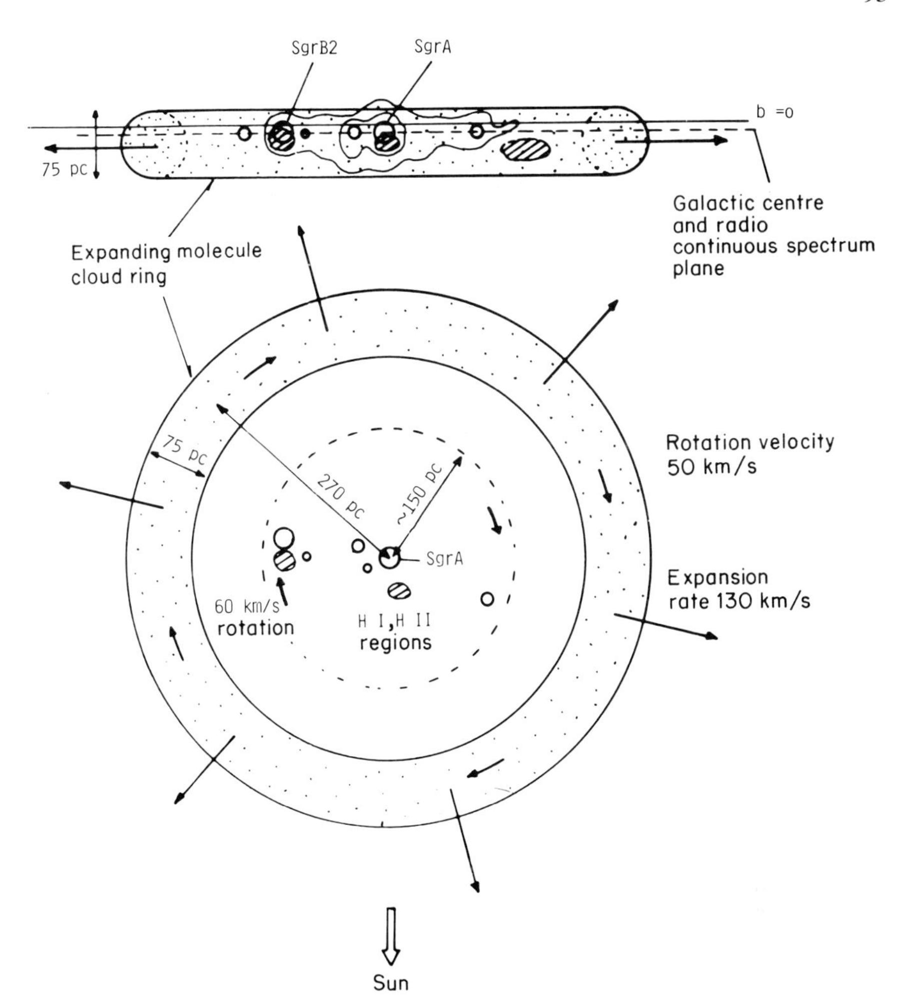

Figure VI.21. The model representing molecule cloud distribution within the galactic nuclear disc, based on many observations, is characterized chiefly by the expanded ring, which rotates at 50 km/s. Apart from the ring, there is probably a whole series of OH clouds and H II regions arranged in circular fashion (OH clouds are represented by broken circles and H II regions by full circles). This inside 'ring' appears to rotate (60 km/s) and contract at the same time. (After Kaifu, Iguchi and Kato)

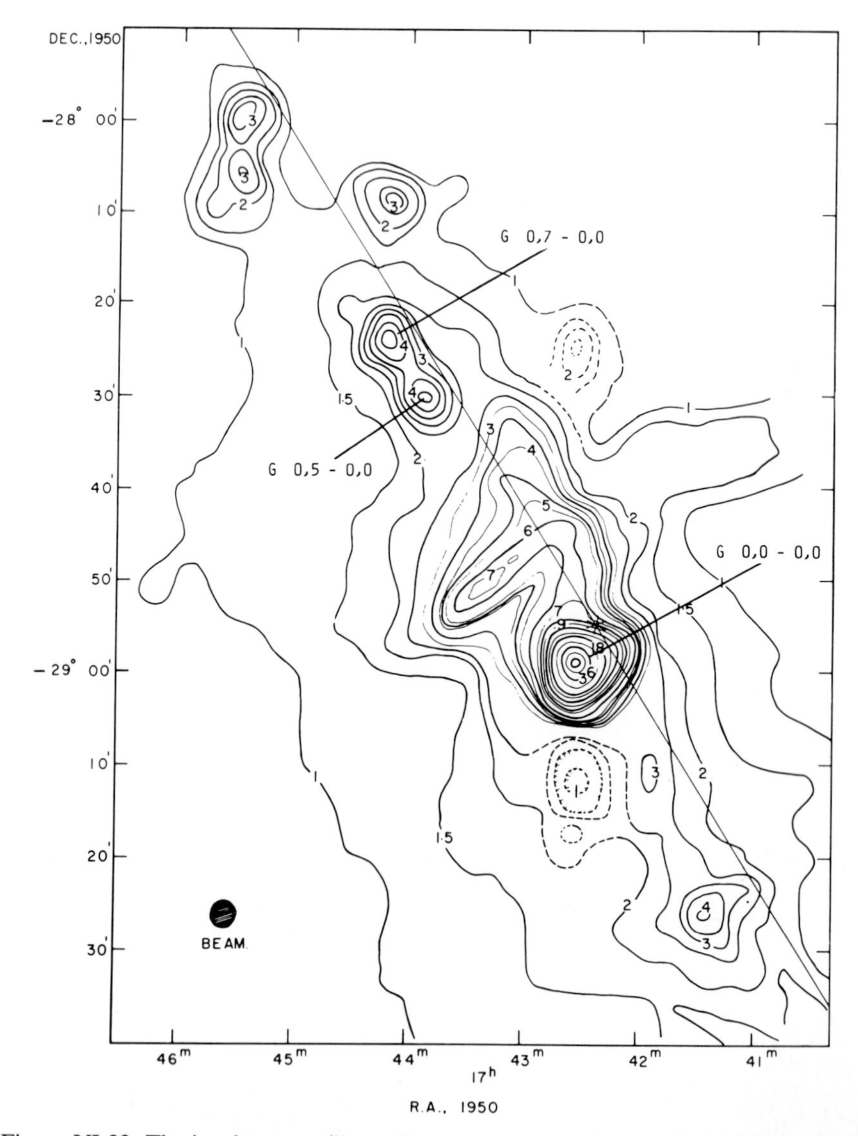

Figure VI.22. The isophot map (lines of equal intensity) of the galactic centre reveals the various radio sources. The map shown here represents radiation at a frequency of 408 MHz. The G 0.0–0.0 or Sgr A source stands out the most. The two sources G 0.7–0.0 or Sgr B2 and G 0.5–0.0 are above this level (After Little)

Figure VI.23. The radiation intensity of radio sources can be shown not only by a contour chart depicting lines of equal intensity, but also by an 'intensity mountain'. Sgr A is clear, but even Sgr B looks just like one of the smaller foothills. Note that the co-ordinate directions are different from those in the previous diagram (After Little)

Figure VI.24. The map of the galactic centre in infrared light at 10 μm reveals nine different distinct radiation sources embedded in a weaker source encompassing them all. The whole section shown here represents the strong radiation source Sgr A

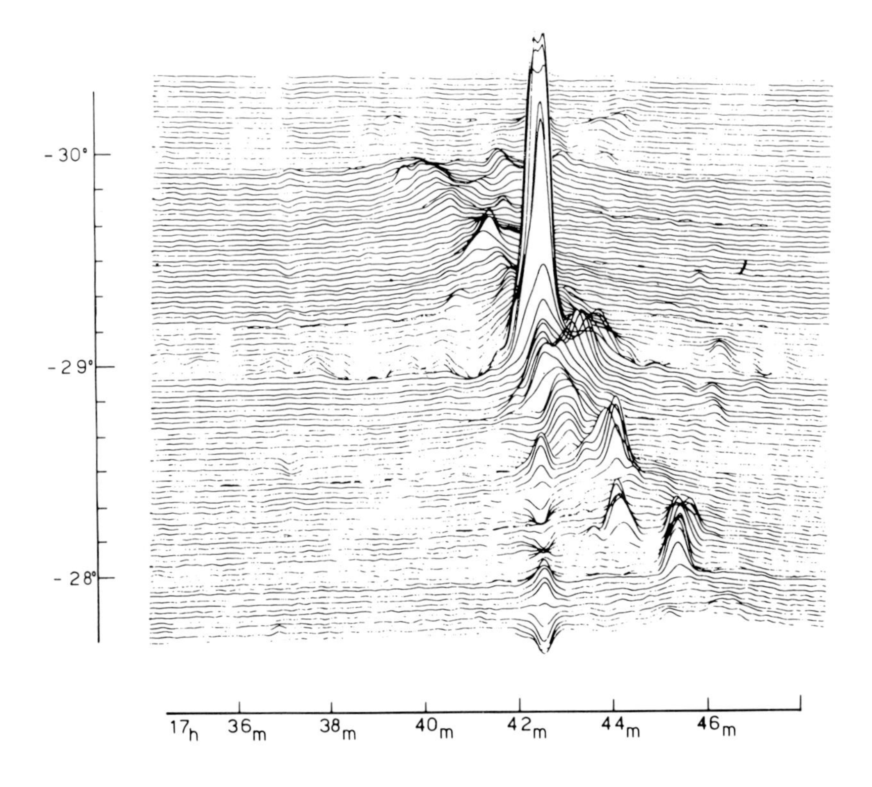
-30°
-29°
-28°
17h 36m 38m 40m 42m 44m 46m

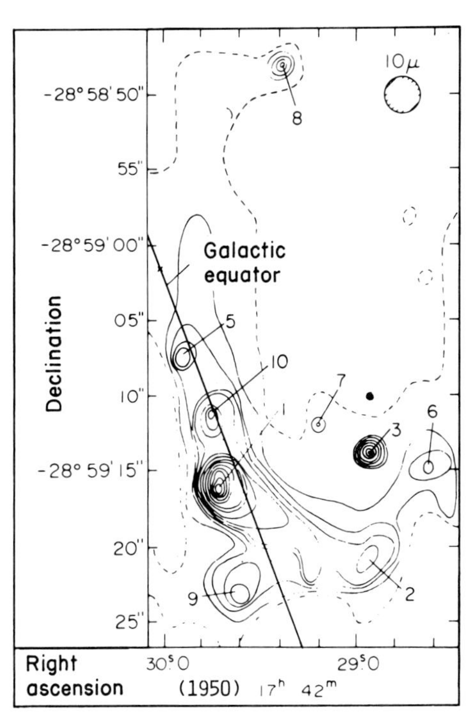
10μ
-28° 58' 50"
55"
-28° 59' 00"
Galactic
equator
05"
10"
-28° 59' 15"
20"
25"
Declination
1
2
3
5
6
7
8
9
10
Right
ascension
30.s0
29.s0
(1950) 17h 42m

VII

The star society

A Glimpse from Without

Now that we have seen how astronomers use the various methods to carry out research into the structure of the stellar system, despite the unfavourable angle of vision from the Sun, we can proceed to take a ride on an imaginary spacecraft, which will take us way beyond the galactic plane to a place from which we can view the whole system from 'above'. The most striking feature from this ideal vantage point would certainly be the spiral structure of the star system. Our Milky Way looks like a slowly rotating catherine-wheel, the spokes of which are the spiral arms which originate in the bright, extremely starry central area. This centre stretches over 6 kpc and is noted for its incredible wealth of stars. The spiral arms go off at a tangent at this point, but not in the form of smooth, narrow lines; they are more likely to be partly dispersed, they seem to be broken up into segments and are sometimes covered with dust filaments. Many of these segments seem to be overlapping, like roof tiles or a shingle structure. The spiral arms wind outwards until they finally break up about 15 kpc from the centre. When viewed more closely, our Sun can be seen about 10 kpc from the centre along the outside edge of one of the arms, 'swimming' along in the disc, together with its colleagues. (More recent research has shown that the Sun is rather less than 10 kpc from the galactic centre. The most accurate measurement to date is 8.7 kpc. However, we shall continue to base our considerations on 10 kpc for the moment, otherwise other distance values would have to be altered accordingly.)

We see a different picture when looking from our spaceship at the side of the system. Far from the spiral structure just seen, the galaxy now resembles a lens with a large, flat brim. The brim is divided up by an irregular, dark stripe. This dark band is the well known interstellar dust, present in a highly concentrated form along the galactic plane. However, something else stands out too: the whole brim is surrounded by a slightly flattened envelope of globular clusters and individual stars, the RR-Lyrae variables in particular. The halo has a diameter of 50 kpc and plays a very minor role in the star rotation system. when

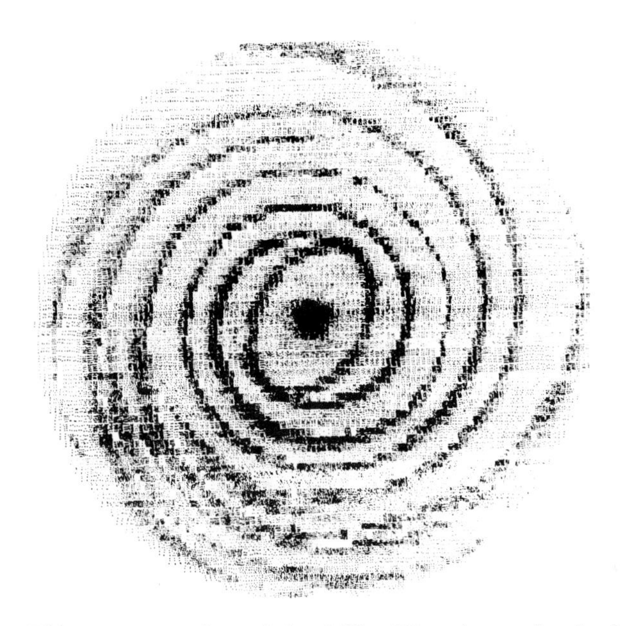

Figure VII.1. This representation of the Milky Way shows clearly the structural components of the galactic centre, galactic disc and spiral arms. As we cannot look at the Milky Way like this in reality, the above is a 'synthetic' picture made by computer. The most significant data observed on a gravitational potential of the galaxy and spiral structure have been used as a basis for the model. Within the Sun's vicinity, the spiral pattern has two arms, but further out it has four. The 3 kpc arm which is near the centre and the 135 km/s arm are seen here as an elliptical ring. (Diagram: Simonson III. Reproduced by permission of Springer Verlag)

the whole is viewed from above, this component cannot be seen as stars and star clusters seem to be projected on to the disc and are therefore not conspicuous.

Blueprint of the System

The highest structural elements of our star system are the galactic centre, the galactic disc with its spiral arms and the galactic halo. They form the skeleton for the definitive construction plan described below.

The nucleus itself is essentially a giant star cluster in which the star density rises to an enormous level as the centre is approached; a figure of 100 million times that in the vicinity of the Sun has been estimated for a distance 0.1 pc away from the centre. These giant star clusters contain a total of more than 10 billion stars and if our Earth were placed in the area it would be as if 100 full moons were shining in the sky at night. In spite of this great density, stellar encounters occur very infrequently, if at all. In addition to the stars, there is also a good deal of gas in the nucleus, mainly hydrogen, which is in the form of a 'nuclear disc' 1–1.5 kpc in diameter and which becomes thicker towards the outside. The gas is atomic on the outside edges, forms a molecular ring further

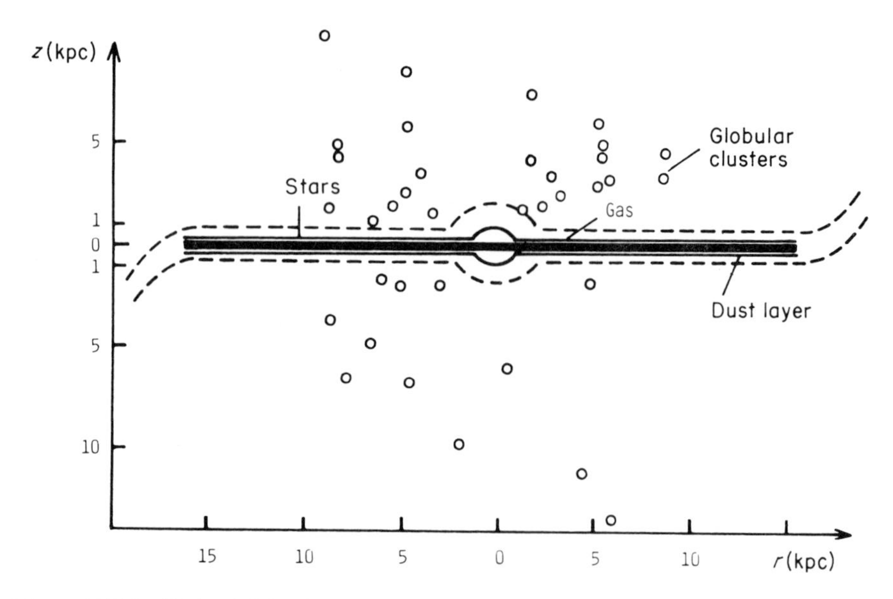

Figure VII.2. If the Milky Way system could be viewed from the side, the most significant characteristics would once again be the 'thick' galactic centre, the relatively flat galactic disc with the light-absorbent band of dust in the middle which curves away at the edges ('hat brim effect') and the large galactic halo of gobular clusters and individual stars. The interstellar matter is concentrated into a layer which is a mere 250 pc thick, whereas the star disc is about 1 kpc thick

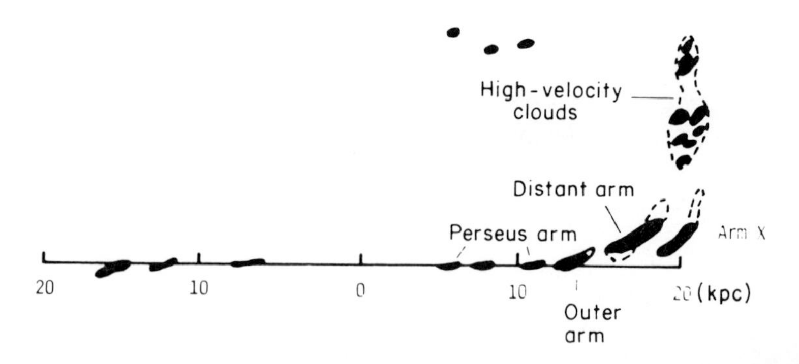

Figure VII.3. The 'hat brim effect' is achieved mainly by the outer spiral arms and clouds with higher radial velocities. (After Davies)

in and is ionized as it approaches the centre. In comparison with the 10 billion stars, the total gas mass of 10 million times that of the Sun is modest, but even so it is about 100 times greater than the amount of dust grains present. The nuclear disc is also a small 'carousel' in itself, revolving very fast and possibly expanding at the same time. The gas in the central region of the disc is ionized by UV radiation from the O or B stars and possibly an excessively large star which may exist in that area, although this latter concept is still purely

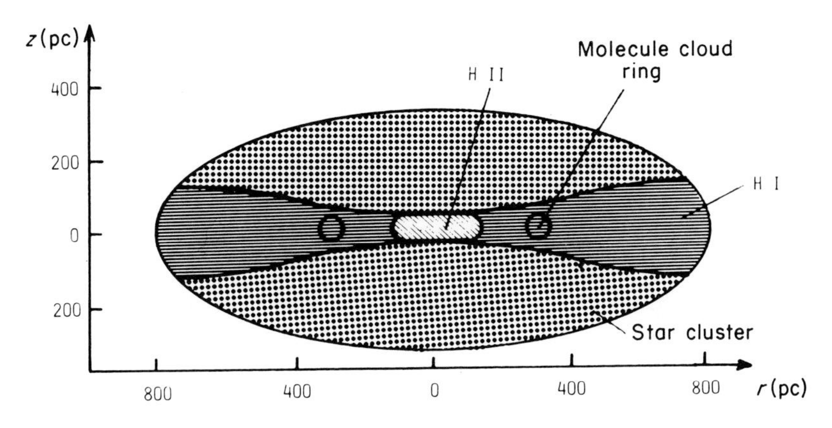

Figure VII.4. The giant star accumulation in the galactic centre is a spheroid covering 1.5 × 0.8 kpc and contains about 10 billion stars. The gas disc incorporated in it contains about 10 million solar masses of gas and is 80 pc thick in the centre and 250 pc thick at the edges. At about 270 pc from the centre of this nuclear disc there is a molecule cloud ring which surrounds the large H II region which is right in the centre. This H II region is formed from the high-energy radiation of young stars

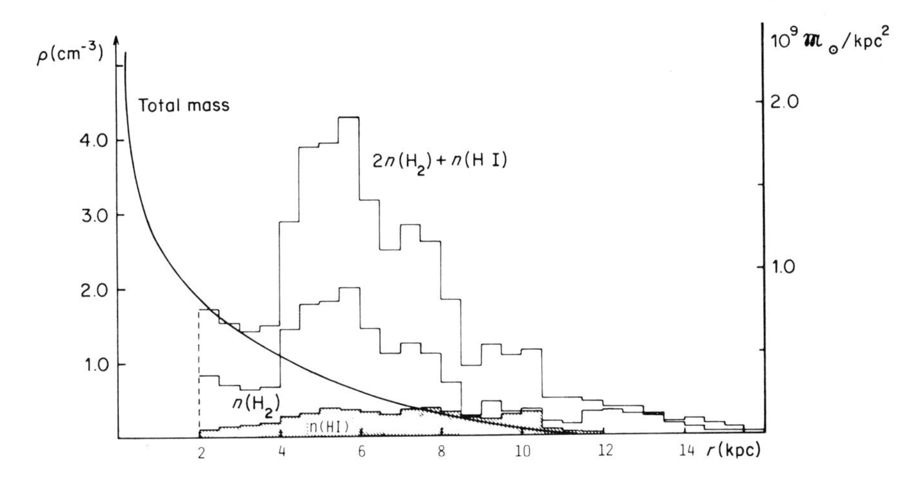

Figure VII.5. Mass density, which is determined mostly by the stars, decreases rapidly with increasing distance from the centre (right-hand scale). The interstellar matter, on the other hand, is concentrated in an expanded ring zone, 4–8 kpc away from the centre (left-hand scale). This applies to molecules in particular; neutral atomic hydrogen is more uniformly distributed (After Gordon and Burton)

theoretical. H II regions of various sizes and the well known radio source Sgr A are also present. The ionized hydrogen zone is enveloped by a molecular cloud ring which also contains the radio source Sgr B. Comparisons with other star systems show that a very special configuration with highly exotic properties is likely to form the very nucleus itself, but at the moment, very little is known about this. The galactic disc, made up essentially of stars, runs outwards from

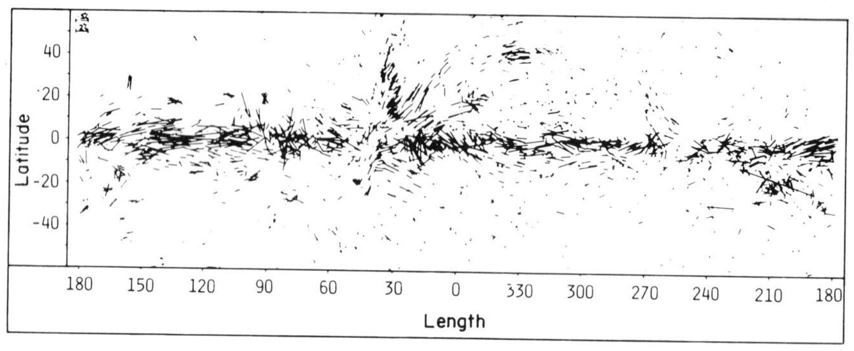

Figure VII.6. In the same way as the magnetic field of a coil can be shown by iron filings on a piece of paper, the interstellar magnetic field can be revealed by polarization of star light. The star light is scattered and polarized on the oblong dust particles aligned by the magnetic field. The position of each small line indicates the direction and the length represents the degree of polarization. When the magnetic field lines run along the spiral arms, then a glimpse into the arm (visual rays are tangential) reveals all possible directions of polarization, but only one alignment is seen when the arms are observed from a vertical position. Polarizations observed point to the Orion arm at l = 85° and 270° and to the Perseus arm at l = 210°. The striking phenomenon occurring at l = 30 could be due to the residue from a Supernova and is in agreement with the radio map (After Ford and Mathewson)

the central region. These stars belong to the disc population, they rotate in circular orbit around the galactic centre and are relatively uniformly distributed—increasing steadily in density towards the centre. The galactic disc is about 1 kpc thick. Scattered in between are the 'high-velocity stars'; they move in an elliptical orbit which is at a much greater inclination to the galactic plane, so every now and then, some of them 'rise up' out of the disc. This galactic disc contains about two thirds of the total mass of the whole system, and includes, of course, interstellar matter which is even more concentrated on the galactic plane than are the stars. Near to the galactic centre we find first of all the 3 kpc arm and the 135 km/s arm and then a comparatively evenly distributed, thin interstellar medium with a gas density of 0.1 hydrogen atoms/cm^3, which forms a large gas disc only 200–250 pc thick. In contrast to the stars, the interstellar gas concentration does not increase towards the centre, apart from in the very nucleus itself. Our observations recorded a density increase of up to an average of 4 particles/cm^3 as the gap approached a distance of 6 kpc away. In fact, the average gas density then drops rapidly towards the outside. Thus, there appears to be a 'gas ring' between 4 and 8 kpc and the space between the centre and this ring seems to have been 'swept clean'. This is particularly so in the case of hydrogen molecule density, whereas atomic hydrogen seems to be more evenly distributed. The galactic gas disc has another distinctive characteristic: the average thickness increases slightly towards the outside, fans out a little at the edges and then bends—upwards on one side and downwards on the other, thus giving the impression of the brim on artists' hats when viewed from the side.

The spiral arm pattern is imprinted on to the relatively smooth galactic disc.

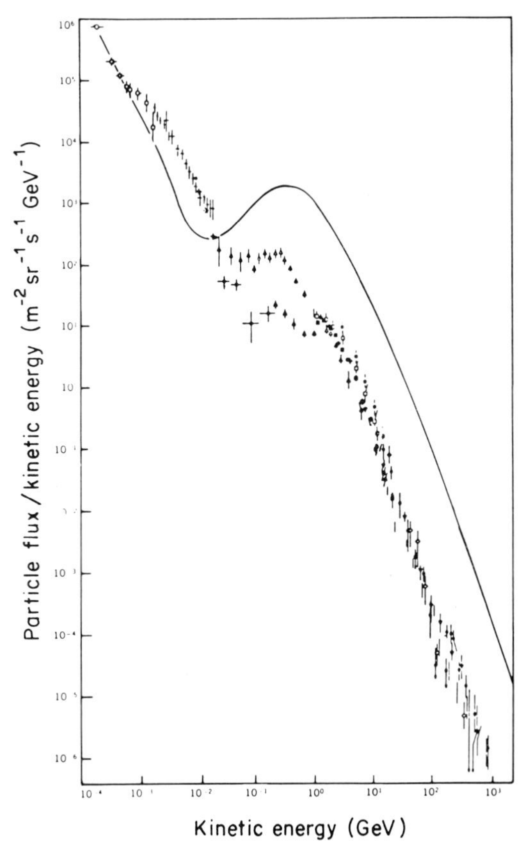

Figure VII.7. This energy spectrum of the protons (line) and electrons (points) in cosmic radiation was measured close to Earth. Below a few GeV the interstellar spectrum is thus influenced by the Sun (relative maximum). The second leg of electron activity between 0.1 and 1 GeV was measured with modified Sun activity (After Meyer, Ramaty and Webber)

The assumption is that two spiral arms start out at a tangent from the central region, less than 4 kpc from the centre and wind out to about 15 kpc away. Rotation of the star system is such that the spiral arms are trailed. A section through a spiral arm would should that the diameter was 1 kpc and that the components were young stars—partly concentrated into associations and open star clusters—interstellar gas and interstellar dust. The spiral arm population is probably not uniformly distributed over the section at all. Interstellar dust and interstellar gas concentrated into cloud formations are the main components along the inside edge of the arm. The average gas density falls very quickly towards the inside at first and then more slowly. The youngest stars are found behind the great accumulation of dust and gas and the slightly older young stars are in the centre and beyond. The average gas density in the spiral arms is about 10 atoms/cm^3, much greater than that found in the inter-arm regions.

Another important structural feature of the Milky Way system is the galactic halo. As already mentioned, it consists predominantly of galactic clusters and

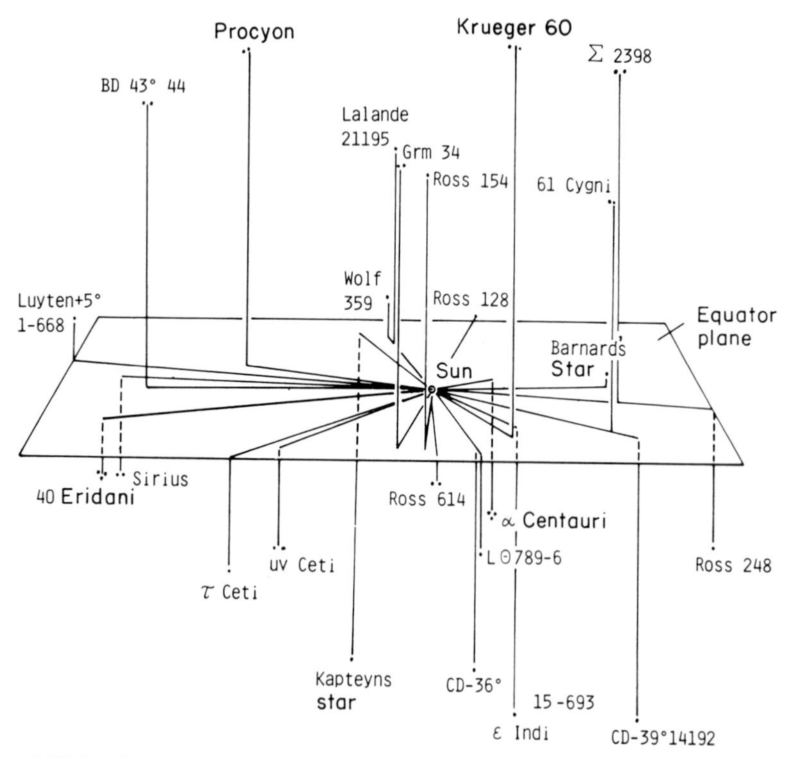

Figure VII.8. There are eight double or multiple systems in the direct vicinity of the Sun (20 of the 30 nearest stars belong to it). The star nearest to us is Proxima Centauri, which belongs to a triple system. The 'fastest' star in the heavens is Barnards arrow star—it has the greatest known proper motion. It is also remarkable from another aspect—it is one of the 'hottest' candidates for the existence of another planetary system

many individual RR-Lyrae stars which surround the disc in the shape of a slightly flattened sphere. The spatial density of the galactic clusters and RR-Lyrae stars increases towards the centre of the whole star system. The total halo mass is estimated as only 20 billion sun masses, making up 10% of the total mass in the overall system. The movement of halo population members is different from that of the disc or spiral arm population. They revolve round the galactic centre in a large, stretched out elliptical orbit which is very steeply inclined to the galactic plane in parts, but do not participate in the general galatic rotation process.

These components are mainly responsible for the outward appearance of the system and justify the concept of the heavens as an ensemble of stars and gas–dust mixture. However,non-material components such as radiation field, magnetic field and cosmic radiation (although strictly this is in fact of a material nature) which are all very closely associated with the stars and the gas–dust mixture, are also included in the blueprint. The whole system is penetrated by the interstellar radiation field, which is made up of the radiation fields of all the individual stars under the 'transformer effect' or interstellar matter. In this

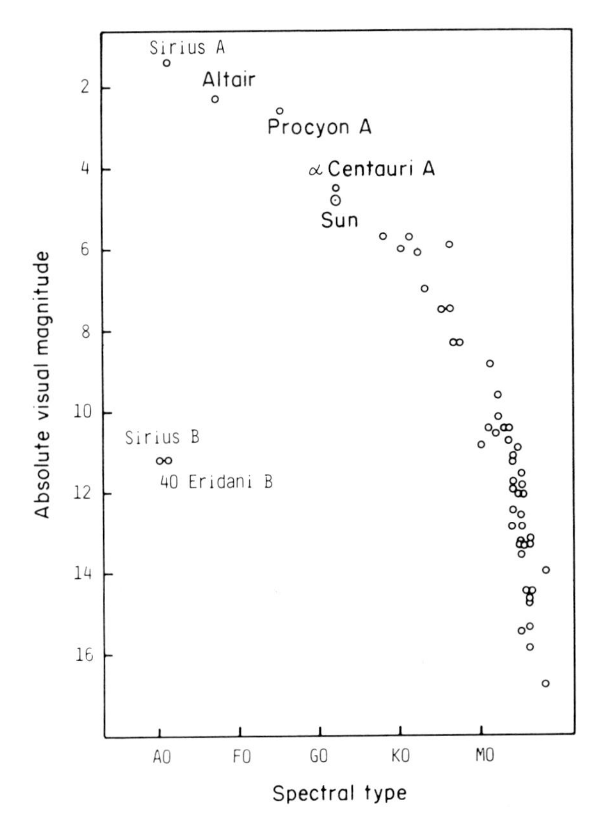

Figure VII.9. The Hertzsprung–Russell diagram for stars in the Sun's vicinity shows that most stars are smaller than the Sun, and that it is the fifth brightest star. The Sun has at least two White Dwarfs and no Red Giants as neighbours (From Van de Kamp)

way, each cubic centimetre contains a certain amount of energy and using the quantum theory to isolate the light into small 'energy parcels', then a cube with 10 cm long sides will contain about 80–100 of these 'energy parcels'. This corresponds to an average energy density of 1×10^{-12} erg/cm^3. At the same time, the star system is also under the influence of an enormous magnetic field, which is much less powerful than that of the Earth but is nevertheless capable of controlling the movement of charged particles. This spatial energy density can also be calculated and works out at about 10^{-13} erg/cm^3 with an average field strength of 2×10^{-6} gauss (gauss is a magnetic induction unit in the c.g.s. system and 0.5 gauss is about the same as the strength of the Earth's magnetic field at the Earth's surface). The shape of the galactic magnetic field has been roughly defined by polarization of star light as when dust particles are arranged 'in rank and file' due to the influence of the magnetic field, the light passing through is polarized in a very characteristic manner. The magnetic fields is closely connected with the galactic plane, is 'coupled' with the interstellar gas and hence is 'fettered' to the spiral arms.

Cosmic radiation is actually of a material nature, being made up of rapidly

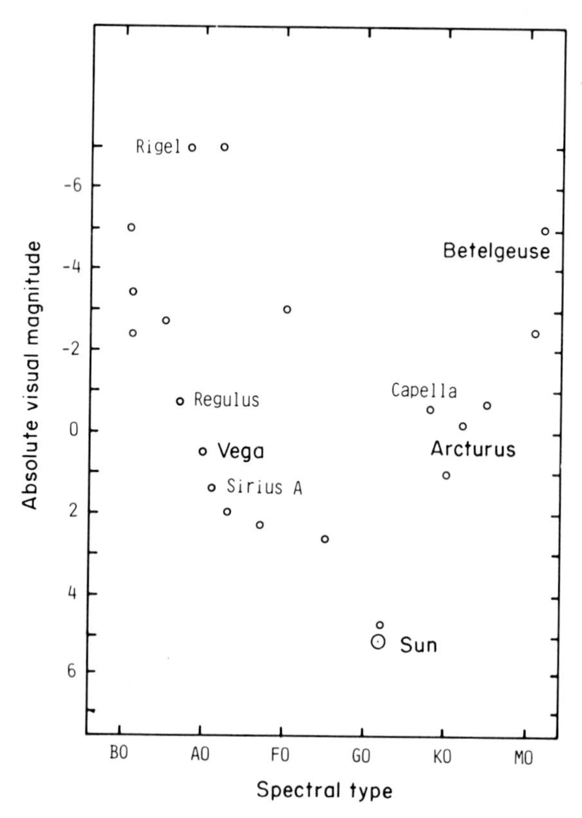

Figure VII.10. The Hertzsprung–Russell diagram for the twenty brightest stars contains six Red Giants and most of the others are very bright and massive stars (often called normal giants). (From Yale Catalogue of Bright Stars)

moving particles. When stars explode (Supernovae) or explosive situations arise in the centre of the Milky Way, 'naked' atomic nuclei, protons, electrons and other elementary particles are released and move exceptionally fast. They 'race' through the whole star system. A rather smaller contribution is made by normal stars which are continuously emitting fast-moving particles into space (this phenomenon is known to us from the Sun as solar wind and is generally known as the 'stellar wind'). Cosmic radiation particles are in part so very closely connected to the magnetic field that they wind themselves around as if it were a rod and thus become trapped in the 'galactic cage'. The diameter of this 'screw' is dependent on the energy and mass of the particles. A single charged particle with the mass of a hydrogen atom and the energy of 1 GeV spirals around the field lines with a radius of 2×10^{-7} pc. However, if the energy level is 10^9 GeV, the radius covers 200 pc and the particle can 'emerge' from the galactic disc gas layer either at the top or the bottom. (An electronvolt, eV is the normal unit of energy used in atomic physics and represents the energy which an electron gains when passing through a voltage difference of 1 V. A gigaelectronvolt, GeV, is 10^9 eV.) The cosmic radiation particles with the

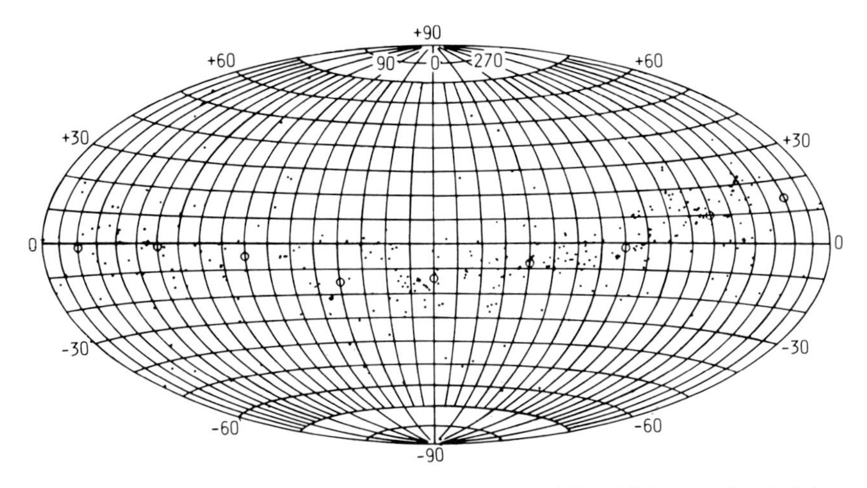

Figure VII.11. The galactic distribution of B stars with a high apparent brightness factor (this means, therefore, young and close stars) forms a wave line. It is interpreted as a small, flat, local star sub-system, called Gould's Belt, and its plane is slightly inclined towards the main Milky Way plane (After Shapley and Cannon)

greatest energy discovered to date have had 10^{21} eV or 10^{12} GeV. Particles with such high energy levels cannot be contained in the 'magnetic galactic cage'. They probably come from other exceptionally active star systems and are simply passing through the Milky Way. The energy content of cosmic radiation amounts to 10^{-12} erg/cm^3 and is thus comparable to that of the radiation field.

Our Near Neighbours—the Sun's Place in the System

Our Sun takes its place with the many billions of other stars as an insignificant nominal member of the galactic disc. In spite of this, it is obviously of great interest to us as it is one of our nearest cosmic neighbours and acts as our 'observation tower'. The Sun is 10 kpc from the galactic centre, 14 pc above the mathematically defined galactic plane and rotates around the centre of the system at a speed of 250 km/s. It belongs to the disc population and lies on the inside edge of a spiral arm. There are an estimated 4500 stars in the immediate vicinity of the Sun, that is, up to about 20 pc away. Most of these are smaller and darker, some are younger and some are older than the Sun. The stars with the greatest degree of absolute brightness in the Sun's atmosphere are Aldebaran (in the Taurus constellation), which is 20 pc away, Arcturus (in the Bootes constellation), only 11 pc away, and Capella (in the Auriga constellation). Our neighbours α Centauri, around which a further star revolves in an 80 year cycle, is only 1.35 pc away. Proxima Centauri is also connected with this double star system and is a further 0.04 pc closer than α Centauri—hence its name (proxima = closest). A comparison of the brightest stars in the system with a list of those which are nearest to us shows that we have only three bright stars, Sirius, α Centauri and Procyon, as neighbours. So, in general, the bright stars are from a much wider area than the less bright ones. However, the true mixture of spectral types and the actual frequency of individual star types can only be assessed by isolating *all* of the stars over a sufficiently large area of space. This challenge sounds easy, but in practice is very difficult to meet. A great deal of time and effort is involved in providing a relatively complete record of all of the stars in a sphere of 5 pc radius alone—a very small astronomical dimension indeed. Some interesting details to emerge from this study of what is probably a typical section of galactic disc are the absence of giant stars, the presence of six White Dwarfs and the fact that our 'weak' Sun is the fifth brightest star in the region. The luminosity function can be calculated from detailed analysis of the Sun's environment (note that neighbouring stars were 'extracted' from the star sea on the basis of the highest levels of proper motion). This function indicates the total number of stars of any specific luminosity range present per pc^3 [or (light year)^3]. This can then be used to record the frequency of individual star types and therefore is of major importance in the field of statistical analysis.

A singular feature emerges when analysing the outer regions of the Sun's atmosphere. The distribution of specific objects in the sphere is often recorded in the form of a galactic coordinate network. If a similar chart is drawn up for the bright stars in spectral class B, then a wavy line emerges around the galactic equator. As this applies to the bright B stars only, i.e. to those which are not so distant, it seems reasonable to assume that the stars in this category form a plane which is at an angle of about 16° to the Milky Way plane. This local system probably extends to well over 700 pc. The astronomer B. A. Gould was the first to test out the theory in 1879 and hence the phenomenon is known as 'Gould's Belt'.

VIII

The Milky Way spiral

The Fate Decreed by Differential Rotation

The spiral structure is an important special characteristic of the Milky Way and many other extragalactic star systems, so an analysis of the forces and mechanisms of these spirals seems a logical step forward.

Let us assume, for the moment, that the spiral arms consist solely of bright, young stars which rotate normally around the galactic centre. Now let us consider a star in the internal part of the spiral and one in the external part. In accordance with Kepler's Law, the star rotating further out will move much more slowly than the inner one. By the time the first has finished one lap, the other will be well into the second. If this 'game' is now thrown open to all of the stars, no matter how near or far from the centre, then it becomes clear that differential rotation winds the spiral arms increasingly tighter and will destroy them completely in less that 100 million years. On the other hand, this same effect can form a spiral structure out of almost any random distribution of bright, young stars in the galactic disc after only very few orbits. Thus differential rotation can contribute greatly to the formation of a spiral, but can also destroy it again; the spiral pattern will not be preserved. Are there perhaps other forces which can help to preserve it from the effects of differential rotation? The galactic magnetic field was the favourite contender for a long time. However, the spiral arms are not composed of young stars alone; interstellar cloud and interstellar intercloud gas are also very important components. Again, this gas is partly ionized by cosmic radiation and hence is coupled with the magnetic field. Thus, the interstellar matter alone could have formed a spiral around the centre and the bright young stars in the spiral arms made from the raw materials entrained would then have had no opportunity to 'slip away' into the general star field. However, more detailed investigation soon showed that the magnetic forces in the Milky Way system are too weak and the galactic magnetic field is not in a position to protect the spiral pattern from the effects of differential rotation.

Attractive Forces made Visible

We have known since the days of Isaac Newton that two spheres at rest will try to approach each other along a direct line of contact. To attempt to calculate the effect of the attractive force of two or three spheres moving at random on another one is much more difficult and the problem of applying the same principles to many thousands of spheres under the influence of mutually attractive forces seems almost impossible. The movement of stars in the galactic disc in the form of a spiral structure represents such a problem and was tackled in the 1920s by the Swedish astronomer B. Lindblad. His conclusions resulted in the First Density Wave Theory, but it remained unrecognized until the 1960s, when it was taken up in America by C. C. Lin and F. H. Shu. This version of the density wave theory is based on the following considerations.

The force of attraction acting on a star in the galactic disc is made up of the various forces of attraction from the many individual stars in the galactic centre and the stars in the galactic disc which move themselves. It can be compared to a cat biting its tail; the stars move within a gravitational field which is built up by the movement of the stars themselves and which is continuously changing. Obviously, interstellar matter is also moving around the galactic centre, but it does not contribute greatly to the attractive force field. At first glance, it seems as though the stars are moving in circular orbit around the centre, but in fact the peculiar motion, which has already been described in relation to the Sun, also has a part to play, This peculiar motion has less influence than the orbital one, but has a significant effect on the dynamics of the system. The combination of the two types of motion can be envisaged as a double circular stellar motion—the star moves around a small circle, the middle point of which itself describes a greater circular path around the centre. If the ratio of angular velocity on the small circle to that on the larger one falls within a certain range, then the result is an increase in star density in some parts of the disc and a decrease in others. Under certain conditions, all points at which density rises or falls may lie on a double-armed spiral emanating from an 8 kpc diameter, evenly distributed central area of the disc and winding outwards to about 16–18 kpc. As a result of stellar motion, slight increases or decreases in density level form a wave pattern through the star gas, which itself obeys the laws of differential rotation and revolves around the centre. Thus, the 'mountain and valley' populations in the wave are subject to constant change—new areas of the disc are taken up, others are left out. The dense mountain and valley areas migrate through the star gas in such a manner that their spiral rotation velocity does not fit with that of the stars. However, the differences in wave density are not yet great enough to reveal any spiral arms and the dense spiral rotates invisibly around the centre. The simple concept of a double rotation circuit falls down on two points. If the difference between star rotation and dense wave rotation were as great as a whole-number multiple of the half-epicycle frequency, then epicycle motion would increase the density right in the centre of a 'dense mountain' and decrease it in the middle of a 'dense valley', causing a

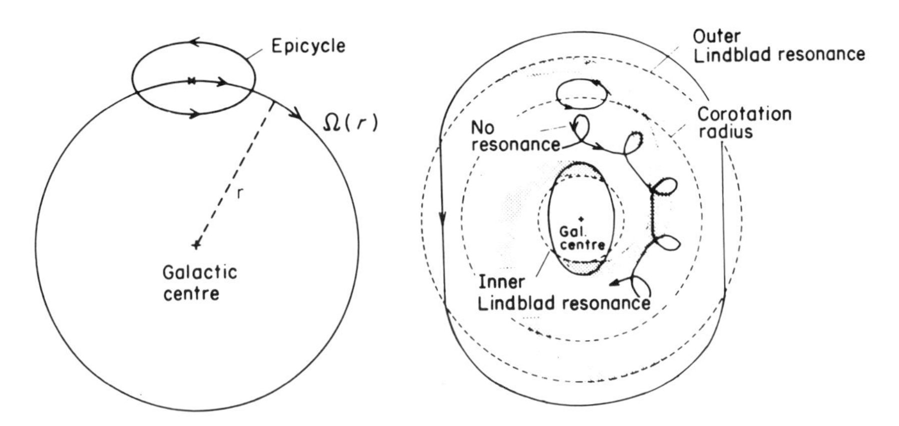

Figure VIII.1. The movement of the stars can be represented in a simplified form as a circular movement around the centre with a second circular or elliptical movement superimposed on the first one (epicycle). An important parameter for composite motion is the ratio between the speeds of rotation of both circles (left). The rotation speed ratio is about 3 at the outer Lindblad resonance, 1 in the corotation circle and −1 at the inner Lindbland resonance. In accordance with the density wave theory, the spiral pattern disappears at the outer Lindblad resonance and the density wave travels as fast as the stars on the corotation circle. A single-figure ratio is used merely for simplification purposes, as in reality resonance occurs when the differential velocity between the stars and the spiral arm pattern and the epicycle velocity is rational, whereby the rotation speed still varies as a function of distance. According to Lin, the inner Lindblad resonance of the Milky Way is 3 kpc away, the corotation circle 16 kpc and the outer Lindblad resonance 20 kpc away from the centre

build-up in amplification, or resonance, to give its proper term. However, as the spiral pattern represents a very small percentage of the general gravitational field of the disc, these resonance points do not correspond to reality. Gravity fields at these points, 4 kpc from the centre for inner Lindblad resonance and 15 kpc for the outer one, must be investigated in more detail. The spiral structure is seen between these two resonance points. The so-called corotation orbit, around which the slowly ending wave moves at the same rate as the stars, is not as far out as the outer Lindblad resonance and the difference between a density arm and a matter arm rotating with the stars disappears.

The spiral density wave pattern around the Sun moves only half as fast as the star gas. Thus, if the Sun is moving at an orbital rate of 250 km/s, the spiral pattern is rotating at only 125 km/s. Despite a slight increase or reduction in star density following the density wave, outwardly the pattern in the galactic disc changes very little at first. The density wave has a much more adverse effect on the gravitational field, which obviously controls both stars and interstellar matter. The gravitational field is altered such that the density wave can 'automatically' penetrate further and thus create the conditions which will ensure its further existence. To be more specific, a rotating, spiral interference field has been superimposed on to the smooth gravitational field of the galactic disc. Star movement is not much affected by the stray field as its peculiar

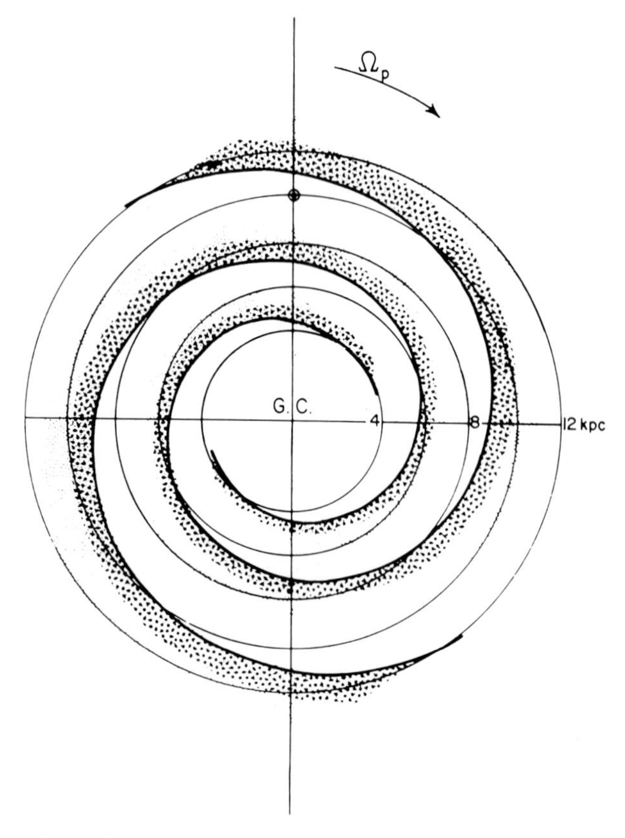

Figure VIII.2. According to model representations, a shock wave is formed on the inside edge of the spiral arms by penetrating gas and this stimulates star production. Thus, young stars and H II areas should be found directly behind the shock wave (large dots). Ω_p represents direction of rotation (After Roberts)

velocity is very high (about 20 km/s) and the gravitational force is stronger the lower is the velocity factor. On the other hand, the interstellar clouds, which have a peculiar velocity of only 10 km/s, are pulled into the 'gravity pit' of the density wave and are considerably slowed down. The disturbed gravity field with the retarded interstellar clouds is still not sufficient to bring the spiral arm into view; the density wave containing the gravitational field pattern would remain invisible if it were not for the fact that new stars are born from the interstellar matter and light up the scenery. Interstellar matter is easily displaced and is highly sensitive to the effects of the gravitational field passing through. If an area of disc comes into the immediate density wave range, then the interstellar matter, which forms into 'lumps' very readily, is sucked into the wave by the powerful forces of attraction. Therefore, the intersellar matter tends to collect together at a gravitational maximum and the gas density is high in comparison with that of the surrounding atmosphere. This increase in density moves along with the wave and, because the latter moves more slowly

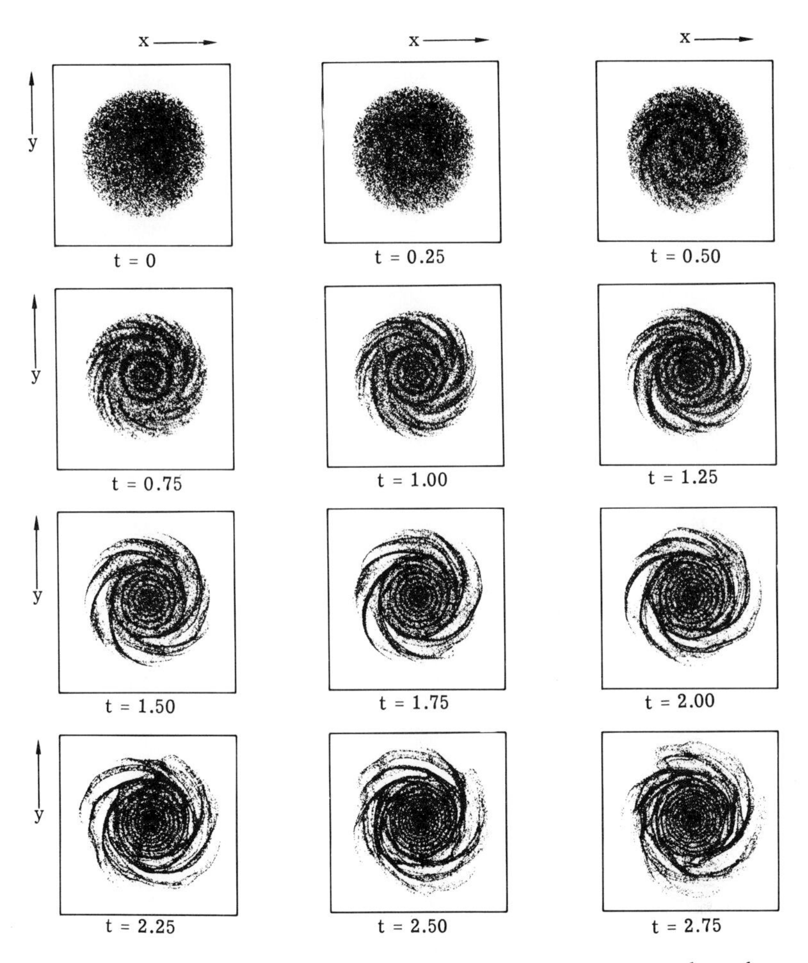

Figure VIII.3. Numerical experiments carried out on large computers show that under certain conditions spiral patterns can be formed in a 'disturbed' rotating disc-shaped star cloud. The figure shows the various stages of development: $t = 1$ represents a period of 250 million years (After Hohl)

than the stars, an observer moving at the same speed as the stars would see that it is rotating in the opposite direction to the general rotation sequence. In this way, the clouds made up of interstellar medium in the galactic disc enter the density wave on the inside. There they bump into existing interstellar matter, causing a sudden rise in gas density. This sudden increase travels through the gas at great speed and is known as a shock front or shock wave. The increase in density in the shock front, which is tied in with the shock wave, is 5–10 times its original level. The drop in density towards the outside is a much slower process. The sudden rise in density when passing the shock front can cause many clouds to become unstable in their own field of gravity. They begin to contract and stars or star families embedded in expanded H II regions are formed. As this birth process takes several million years—and as protostars, they remain invisible—the shock front has moved on several hundred parsecs further by then, so that all newly formed stars, star associations and H II areas in the general star field do not come to light until the shock wave has passed by. In spite of this, the peculiar motion which they inherited at birth prevents them from becoming lost in the general star field at this stage. Their brightness factor and the typical way in which they are arranged outline the spiral arm structure of the Milky Way. It reacts rather differently with the interstellar dust which has particularly close links with neutral hydrogen gas. As the H I density in the shock front is greatest along the inside edges of the shock wave, the dust grain density is also at its highest at this point and the dark clouds and filaments should be on the inside of the spiral arms. Observation has roughly confirmed the theory that the spiral arm population consists of decidely young objects. However, these latter do not remain in the spiral arm for ever. As it passes along further, it leaves the stars, star associations and star clusters which have reached 'maturity' behind in the general star field and becomes conspicuous itself again by the emergence of new 'flashing' stars.

The difference between orbital velocity and density wave velocity, which is the same as the gas impact speed, is greater in the region of the inner Linblad resonance than it is further outwards and it drops to zero for the corotation orbit. The greater the impact velocity, the higher is the density in the shock front and the better are the conditions for star evolution. This is why there are considerably more H II regions further in, at around 4 kpc from the centre, than there are further out, although there is sufficient hydrogen present even in the corotation orbit.

The dynamic reaction of interstellar matter as the density wave passes through is extremely interesting. As it enters the density wave area it has one tangential and one radial movement component. However, the retardation effect destroys the radial one almost completely, whereas the tangential one remains essentially intact and interstellar cloud and newly formed stars leave the density wave almost at a tangent.

One unsolved problem arising from the theory outlined above is the so-called dissipation of energy. Continual collision of interstellar matter from the inter-arm region on the inside of the spiral arms causes wholesale destruction of part of the outward velocity component. This means that energy is transported radially and this energy must obviously come from somewhere. The question is where? There is no known answer at present. Does the enigmatic nucleus of the star system provide the source of energy? Or does the whole star system have an outside 'supply' from which highly dilute hydrogen gas is drawn in? The search for a solution continues.

The density wave concept as an explanation for spiral pattern is based on the assumption that the spiral structure of the gravitational field is already there and attempts (very successfully) to give physically viable solutions as to why the structure is preserved despite differential rotation. But where does the impact that 'sets the ball rolling' come from? This question represents another of the

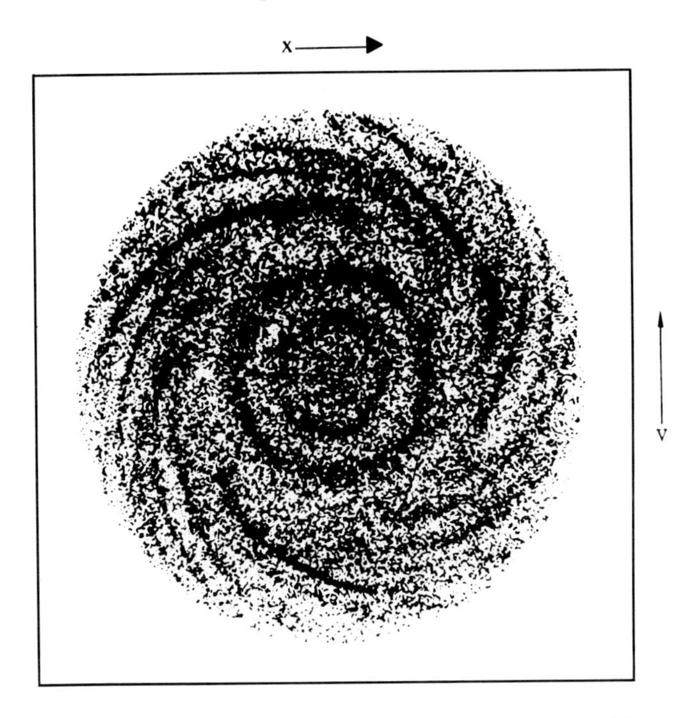

Figure VIII.4. An enlargement of the picture for $t = 0.75$ (i.e. a period of 187.5 million years) gives an even clearer image of the spiral structures. However, a ring has formed around the actual centre (after Hohl)

unsolved problems embraced by this theory and, in fact, a single stimulus would probably not be enough because, left to its own resources, a density wave pattern would dissipate and vanish over the course of a billion years. Three possible explanations are being considered by astronomers at present. Firstly, locally confined, random changes in gas density could take over the role of stimulus mechanism. Secondly, some astronomers point emphatically to the explosion and centrifugal processes which occur in the central part of the star system and which could also be the cause of the density wave phenomenon. Finally, another group believes that the tidal forces produced when two star systems pass close together provide the conditions under which the spiral pattern is formed. However, no precise details have emerged to date.

Although the density wave theory of spiral pattern cannot possibly be confirmed by observation (thus, the local or Orion arm, for example, does not readily fit this concept), 'numerical' experiments related to spiral structure have been made on large computers over the last few years. These show that our ideas are not totally divorced from reality. With the help of a few 'tricks', the movement of about 100,000 mass points (stars), moving about in their self-generated gravitational field, was calculated in detail. When an additional force field was introduced into the whole mass point complex—and this, in fact, only conforms in part to conditions actually prevailing in the star system—then a spiral structure with arms representing density waves did emerge. Thus, these 'computer games' could be regarded, with some reservation, as a qualitative confirmation of the density wave theory.

IX

Brothers, sisters and children (?) of the Milky Way

Two Striking Phenomena in the Southern Hemisphere

The seafarer Fernando Magellan noted variations in the heavens during the course of his travels across the Earth's oceans. In 1519 he saw two bright spots in the southern constellations Dorado and Tucana, which remained well lit even by moonlight and were therefore ofen used for navigational purposes. These two stars, which have been known ever since as the Large Magellanic Cloud and the Small Magellanic Cloud, are not far from the southern celestial pole and are seen through relatively small telescopes as gigantic star clouds. They became the subject of some attention after John Herschel published his survey of the southern sky. Following in the footsteps of his celebrated father, he had begun to search the heavens systematically for nebulae, star clusters and double stars. He worked with exceptional diligence and in eight years had discovered 2306 neublae and star clusters and 3347 double stars in the northern sky. 'Spurred on by the enormous significance of the discoveries and the magical nature of the emerging objects', he determined to carry out a similar survey of the southern sky. Over a period of four years he isolated 1708 nebulae and 2102 double stars from an observatory at the foot of Table Mountain just outside Cape Town. He laid special emphasis on observation of the Large and Small Magellanic Clouds. Herschel counted 278 various star clusters and nebulae in the large cloud and another 50 a little to one side. He catalogued 919 objects in this cosmic structure. The profusion of nebulae and star clusters led him to the conclusion that 'the clouds should be considered as a special kind of system which has no analogy in our hemisphere'. His comment on the relative star void in the surrounding celestial areas, particularly noticeable around the Small Magellanic Cloud, was 'a desert, completely enclosing an oasis in full bloom'. However, his concept of the two Magellanic Clouds being original star systems did not gain favour until later.

Both Magellanic Clouds in the southern sky are the very closest star societies to us outside our own star system and so they are related to our Milky Way,

Figure IX.1. The Large Magellanic Cloud is about 50 kpc away from us and has a diameter of 11 kpc. The lamella in the middle is especially clear. It is a substantial structure on the same scale as the spiral arms of our Galaxy; stars are always being drawn into and ejected from the lamella. In contrast to the Milky Way system, the spiral structure, starting at the bright complex (30 Doradus) to the left above the lamella, is only hinted at. (Photograph: Mt. Stromlo Observatory, Australia)

despite their different structure. The Large Magellanic Cloud is about 50 kpc and the Small Magellanic Cloud is 65 kpc away from us. Although this seems an enormous distance, they are in fact 'on our own doorstep' in comparison with other cosmic islands. As a result, they seem to extend over a large area of the sky: the Large Magellanic Cloud has an apparent diameter of 11.8° and that of the Small Magellanic Cloud is 4.2°. The actual sizes as a function of the given distances work out to be 11.8° and 4.6 kpc, respectively. Thus, in comparison with the Milky Way, these systems are relatively small. However, the shape and the size of the system is different from ours. Components such as our disc with its spiral structure and the dense nucleus are not present at all in the Magellanic clouds. They are much more irregular in construction and only the Large Magellanic Cloud shows any sign of a spiral pattern at all, in the form of a barred spiral. One property of the clouds is of special interest to astrophysicists: both stars contain a great mass of interstellar matter. About one tenth the mass of the Large Magellanic Cloud consists of interstellar gas and dust and about one third the mass of the Small Magellanic Cloud is taken up by the

Figure IX.2. The Small Magellanic Cloud is only 4.6 kpc in size and shows very little sign of a spiral structure. The two Magellanic Clouds are companions to the Milky Way and probably rotate around it in an inclined orbit. (Photograph: Mt. Stromlo Observatory, Australia)

same. This interstellar matter is clearly visible in the form of light and dark nebulae and is also picked up on radio frequency waves. In fact, radio radiation has uncovered some properties of Magellanic Cloud movement. Apparently, both clouds move around their common centre of mass over a period of time, which corresponds approximately to the rotation time around their own axis. In this way, the clouds always 'face' each other the same way and rotate, rather like a dumb-bell with an invisible pole in between.

Study of the two Magellanic Clouds is an interesting aspect of astronomy itself, but it also helps towards a better understanding of the construction and evolution of our own star system. We can view these two small neighbouring galaxies from the outside and this gives a better idea of the distribution of stars, star clusters and nebulae than we get from within the Milky Way. Moreover, the great problems caused by absorption as a result of interstellar dust, which make distance calculations so tedious in the Milky Way, no longer apply. In comparison with the distance between the Sun and the Magellanic Clouds, the 'small' distances within the Clouds themselves are of very little significance, so that all Cloud objects can be regarded as the same distance away for all practical purposes. In this way, factors such as the period–luminosity relation of Cepheides-type variable stars can be accurately fixed. The same applies to tests on star clusters and H II regions. More than 1600 star clusters and 500 H II regions have been isolated in the Large Magellanic Cloud alone, figures way above those for the much larger Milky Way system. Last but not least, the chemical composition of stellar and interstellar matter in the Magellanic Clouds can be calculated spectroscopically and compared with that of the Milky Way. This makes a very important contribution to research into the structure and evolution of the star system. Hence it is hardly surprising that many astronomers have devoted much time to this aspect over the last few years and large telescopes have been set up in the southern hemisphere specifically for this purpose.

The Andromeda Nebula—the Milky Way's Twin

It would be extremely surprising if our small Earth and its Sun belonged to a system which was unique in the vast regions of space. Indeed, this formed the basis of the argument that the distant nebulae were cosmic islands, first proposed by Immanuel Kant and a very popular belief during the last century. However, lack of detail about the design and structure of the Milky Way prevented any confirmation of the theory, The fact that the Milky Way did have brothers and sisters more or less similar to it was proved at the beginning of this century and tests on the well known Andromeda Nebula were a main contributory factor. Its properties were directly comparable to those of the Milky Way. During the 1920s, the American E. P. Hubble succeeded in splitting up the outer edges of the Andromeda Nebula into individual stars and giving a rough estimate of distance. In 1944, Walter Baade defined the bright central area as dense accumulations of individual stars. Since that time, this Nebula—or star system to use the preferred term now— has become a prime example for comparison with our Milky Way system, mainly because it is so similar and it is only a 'short' distance away.

Just as there is a disc plane which is often used as a reference in the Milky Way system, there is a similar 'Andromeda plane' in the Andromeda system. We cannot see it very well as we look at it at an angle of only 14°. This has disadvantages, but it also has advantages. The small angle of view is not good for research into the spiral structure and nuclear area of the Andromeda system, but rotation and interstellar dust distribution can be readily and accurately defined. So, what are the most important properties of our Milky Way 'twin'?

Like the Milky Way, the Andromeda system consists of a dense central area and a disc which displays a spiral pattern and two arms which wind themselves around the nuclear area about three times each. The arms are formed by an accumulation of star associations and H II areas. As the angle of view to the Andromeda plane is so small, the light absorbent effect reveals the interstellar dust concentrated in the disc as a dark band and the whole system is surrounded by a halo of globular clusters just as ours is. The central area of the system with its star-like centre can be seen in the optical range from the outside. As a result,

Figure IX.3. Our nearest spiral nebula is the star system M 31 in the Andromeda constellation, about 700 kpc away. It is 33 kpc in diameter and about the size of the Milky Way system. The two smaller star systems NGC 205 and M 32 are companions to the Andromeda system. The diameter of the elliptical NGC 205 is 2.4 kpc, whereas M 32 is about 3 times smaller. (Photograph: Mt. Palomar Observatory)

the nucleus has been photographed, radiation from the central area recorded in the near infrared frequency range and star density calculations then worked out accordingly.

The structure of the Andromeda system is so similar to that of the Milky Way, both generally and in many of its details, that a method for research into our own galaxy has been derived from it. We cannot determine star density in the central area of our system direct, but infrared radiation can be measured. Both star density *and* infrared radiation can be measured in our neighbouring spiral so, in view of the general similarity of the two systems, the galactic star density in the central area can be obtained by comparing it with the ratio between star density and infrared radiation in the Andromeda spiral. The Andromeda nucleus itself is about 8 pc in diameter and contains 10 million sun masses. More detailed analysis of the spectra of Andromeda stars, gas nebulae and radio frequency waves disclosed differential rotation and showed that the

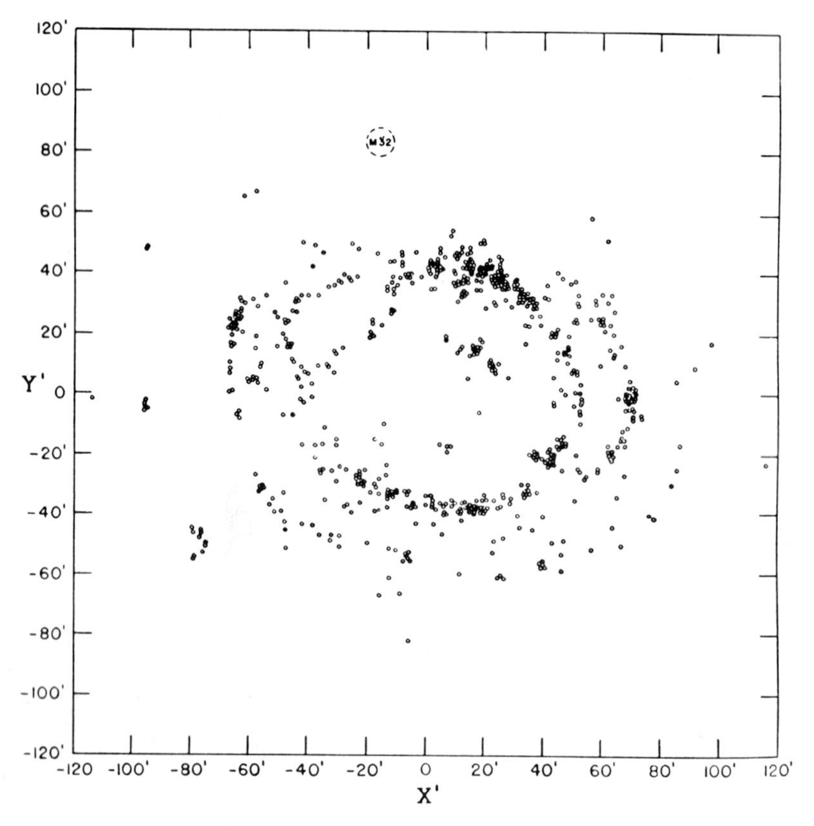

Figure IX.4. The position of the bright emission nebula in the Andromeda system discloses the typical spiral structure. The plan view shown in the figure is obtained by taking a flat viewing angle. The density wave theory applies to the Andromeda system in the same way as it does to the Milky Way system, and emission nebulae are arranged along the spiral arms accordingly (After Arp, 1964)

system is moving towards us at 300 km/s. The orbit of a star 10 kpc away from the Andromeda centre lasts about 250 million years which corresponds to the Sun's orbit period around the galactic centre.

One feature of the Milky Way system is its two satellite moons—in the shape of the Magellanic Clouds. The Andromeda system also has the 'luxury' of two smaller moons in the form of the small galaxies NGC 205 and M 32 (the first is listed in the catalogue of the English astronomer Dreyer under number 205 and the second in that of the French astronomer Messier under number 32). However, these two are different in appearance from the Magellanic Clouds. They have a much more regular elliptical form. Very little is known about the movement of the two satellite galaxies. Also we do not know whether or not they were captured by chance, formed at the same time as the main system or even possibly flung out following a gigantic explosion in the central area and are therefore true 'children of a mother galaxy'.

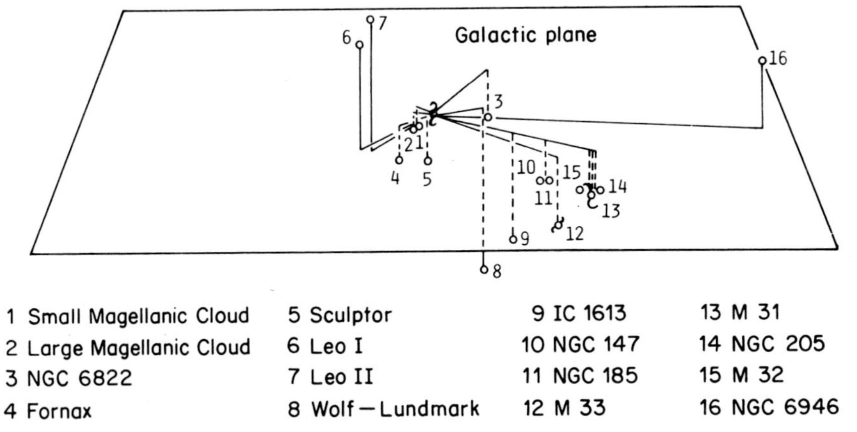

Figure IX.5. There are 18 relatively fixed members and 12 candidates in the 'Local Group'–a system of star systems. The two biggest are the Milky Way and the Andromeda systems, with several smaller ones grouped around them. The two Magellanic clouds belong to the Milky Way, or galactic sub-group. The Andromeda sub-group consists of the Andromeda system with its companions NGC 205 and M 32, and also NGC 185 and NGC 147. Of the 18 fixed members, 3 are spiral, 10 elliptical and 5 irregular in shape (in spite of its vague spiral structure, the Great Magellanic Cloud is regarded as irregular)

The Universal Hierarchy

During the eighteenth century, Johann Heinrich Lambert put forward the theory that all heavenly bodies are part of a larger and more comprehensive system, which then forms part of another one and so on *ad infinitum*. The first stages of this are easily confirmed in our own cosmic environment: the Moon rotates around the Earth, the Earth–Moon system and the other planets go round the Sun and the Sun system together with many millions of other 'suns' orbits the Milky Way centre. In Lambert's time, the latter stage was unknown and the 'central sun' of the Milky Way system was searched for in vain for many years. Nowadays, we know that the individual stages in the hierarchy are not the same, but show a great degree of diversity. The Sun system as a whole is much more complicated and indeed such more than just an expanded form of the Earth–Moon arrangement. This becomes easier to understand if we switch from sun to star system. As we now know a good deal about our star system, the neighbouring Magellanic Clouds and the Andromeda structure, we should ask ourselves whether the time has arrived to take another step up Lambert's hierarchical ladder. Is the Milky Way 'only' a member of an even more powerful system? With some reservation, we would say that the answer to this question is affirmative. Caution is essential, though, because the next system up, to which the Milky Way belongs, is comparatively modest in size and actually has no strict regime. This system, the so-called 'local group', seems much more like a random and colourful company of several star systems 'thrown together'. The most important and largest members of this group are the Andromeda and the Milky

Figure IX.6. The third spiral of the local group, the star system M 33, is in the Triangulum constellation. This system has much more loosely structured arms than the Andromeda system and is also considerably smaller than the two other spiral systems. It is a mere 18 kpc in diameter. The M 33 system is 720 kpc away from us. (Photograph: Mt. Palomar Observatory)

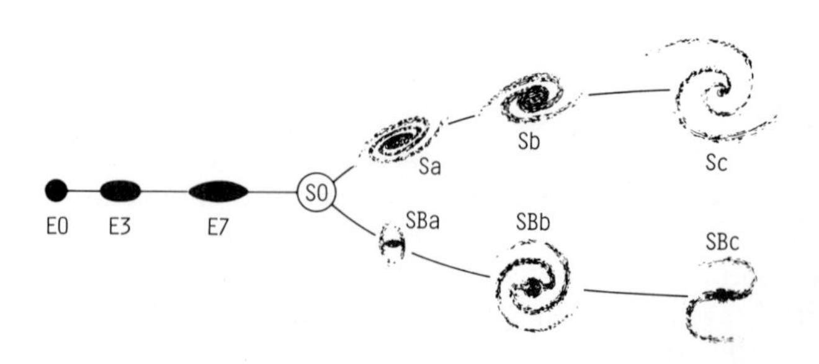

Figure IX.7. The most widely used method of classificatiom for extra-galactic star systems was devised by the American E. P. Hubble. He divided the star systems into four groups: elliptical, spindle-shaped, spiral and irregular. These main groups were then further sub-divided. The most remarkable feature is that there are two types of spiral: the normal sort which has arms tangential to the centre and the barred spiral which starts from lamellae. Normal spirals occur much more frequently than the barred ones

Way systems. At the same time they are a little like the spatial poles of the local system as most of the other much smaller members group themselves around the two largest ones. The outside impression of all this is that the local group consists essentially of two large star families represented by the Milky Way and Andromeda groups. We are already familiar with the two Magellanic Clouds from the Milky Way. In addition, it has a further seven smaller galaxies, plus eight further candidates which have not yet qualified for membership. The Andromeda family consists of NGC 205 and M 32 plus three further galaxies. In addition, six 'outsiders' have been found scattered about in the space of the Local Group, bringing the total membership of this system of systems to 18, with more than 10 candidates on the waiting list.

We are still unaware of how far these various star systems are involved in a common, systematic process of motion. The systematic character of the group is, therefore, expressed solely on a 'neighbourhood' basis, by the fact that they are all 'chained together' by the forces of gravity. No galaxy moves fast enough to overcome the attractive force of the others and really 'tear itself away' for ever. We say that the system is gravitationally bound.

The satellite system of both the Milky Way and the Andromeda structures show that star systems can differ from each other greatly both in size and shape. E. P. Hubble was one of the first to study systematically the nearer, easily visible star systems outside the local group and to classify them. According to his method of classification, the four main categories are the elliptic, spindle-shaped, spiral and irregular groups and each class can then be subdivided further. In addition, there are two types of spiral systems: those in which the spiral arms depart at a tangent from the nucleus and those in which the spiral starts from a bar-like concentration in the central area. Local group population (accepted members and candidates) is, therefore, made up of the two great spirals, the Milky Way and the Andromeda systems, the smaller spiral M 33, ten smaller irregular systems, 15 elliptical dwarf systems and the two elliptical systems, NGC 205 and M 32. The diameter of our star-system system, or galactic cluster to use a better term, is about 2 million pc.

X

Formation and evolution of the star system

Many tests have been carried out to establish the age of the Sun: we know that it is at least 5 billion years old and that its energy supply will last for 'only' a further 5 billion years. However, as the most important part of stellar population, the stars, are born, evolve and eventually 'die', then it follows that we must want to know how these processes affect the system as a whole, whether and how the Milky Way was formed and its prospects for the future. Some results of recent cosmological research indicate that the Milky Way was formed and evolved as one galaxy amongst many. Today, there are very good grounds for taking the starting point of our tangible world as around 15 billion years ago. We do not know what it was like before this time, only that the world must have been very exotic. The rapid expansion of the world began with a 'big bang' and during the course of this process of expansion, which has continued and is continuing even now, the star systems including our Milky Way were formed. Thus an upper age limit can be put on the star system dating from this 'world creation' period; the Milky Way cannot be more than 15 billion years old and it must have a history. However, the story is very difficult to check as it took an extremely long time to write, in comparison with our life on Earth. However, let us see what changes occur as the Milky Way revolves just once around its own orbit. We call this period of 250 million years a cosmic or galactic year. During this time, many stars will form and many others will 'die'—and the wasteful O and B stars will not yet be one galactic year old. The star clusters will change—loosely structured configurations will disintegrate, others will change shape and interstellar positions will be different. Maybe there will be one or two explosions in the galactic nucleus itself. Even so, a distant observer will notice scarcely any change, even after a cosmic year has elapsed. Despite this, the Milky Way is not without its history: firstly, let us examine its 'diary' to establish the motive forces behind the whole evolutionary process.

'Development Motors' of the Star System

About 90% of matter in the star system is contained in the stars themselves. They are made out of gas and dust grains, 'live' by releasing huge amounts of internal energy by nuclear fusion and will eventually 'die'. However, they are not all of the same age and their lifespans vary. Observations have shown that stars are being formed from interstellar matter even today. However, the interstellar matter present in the star system now is different from that which was around during the 'very first hours', and this obviously affects recently formed stars and those about to be born. What makes this so? The life of a star consists predominantly in building up higher elements by nuclear fusion, during the course of which important life-giving energy is released. Under favourable conditions, the elements can continue to be built up in this way until iron is formed, then the process stops as the star does not have the energy required to produce heavier elements. Thus, when a star has come to the end of its period of nuclear fusion, it has reached 'pensionable age' and becomes a White Dwarf, neutron star or even a 'Black Hole'. It can only be a White Dwarf or neutron star if it is smaller than 1.5 sun masses, so in some instances the path to a 'star pension' must be strewn with a considerable amount of discharged mass. Planetary nebulae represent this type of discharged gas cloud—the central star 'wants' to become a White Dwarf. Another form of mass release is the Supernova—a star explosion in which most of the star mass is flung into space and sometimes leave behind a neutron star as a relic. During the course of a Supernova explosion, higher elements than iron are formed as a result of various fast nuclear reactions in the rapidly expanding envelope—the extra energy required being supplied by explosions. There are also other less stimulating forms of mass discharge: stellar wind is a more generalized form of the well known solar wind and is made up of a stream of particles moving away from the star. Through this process, the Sun flings about 10^{14} tonnes of matter into space every year. This represents a very small part of its total mass but many supergiants can bring the figure up to three hundred-thousandths of a solar mass per year, or 100 million times the activity of the Sun. Another mass-release process is the Nova outburst in which a thousandth to one ten-thousandth of a solar mass is released each year. These explosive events obviously have an effect on the existing interstellar matter; they heat up the gas and whirl the interstellar clouds around; in short, they affect the dynamics of interstellar matter.

All of these mass release processes supply the interstellar matter with material which has already been 'well cooked' in the star and thus is well endowed with heavier elements. As a result, newly formed stars contain a greater percentage of these heavier elements as part of their make-up. Star evolution influences the 'life' of the star system by changing the composition of part of the interstellar matter component, by changing the movement patterns of interstellar cloud and by changing the properties of star relics; the cooling White Dwarfs and neutron stars and the Black Holes in particular no longer make any useful contribution to the radiation field and the very existence of these is obvious only from their forces of attraction. Star evolution is *one* factor in the development of the star system.

The external appearance of the Milky Way, in common with the other star systems, is dictated by the stars and their movements. The stars are arranged in a disc and move in large circular orbits around the centre. However, in contrast to the Sun's system, the gravitational field which controls the stars is in this case produced by the stars themselves and is thus not a fixed characteristic from the start. Star movement reacts upon the gravitational field, which in its turn controls the former. Moreover, star movement is not without its disturbances. Despite the low density in comparison with that of the Earth, stellar encounters both between stars and with interstellar cloud do affect movement patterns. However, these 'star collisions' are not real collisions at all but occur when two objects are passing each other at more or less close range. Star density is much too low for real encounters—estimates are that such an event, which would have catastrophic results for the stars involved, is likely to occur only once in a billion cosmic years in the vicinity of the Sun. Even the likelihood of a star passing closer than the distance from the planet Neptune can be excluded, for all practical purposes. On the other hand, stars passing 0.5–3 pc away cannot be ignored. When this occurs, the stars 'shake hands briefly' owing to force of gravity, and both pull out slightly from their orbital path. The deviation when saying 'good-day' is no more than one arc minute, but it happens so frequently that it adds up over the course of a cosmic year and the total then corresponds to a meeting over a minimal range of 100 astronomical units, which is a considerable amount. Even the mass discharged during star evolution can affect star movement. A reduction in the angular momentum or the 'eruption' from a double star system may be a result of mass loss. When transferred to the whole star system, all of these processes have a direct influence on the dynamic factors responsible for the formation of the spiral pattern, differential rotation and hence mass distribution throughout. Stellar motion or dynamics is, therefore, a second motor for assisting the development of the star system. Obviously, gravitation represents the 'sinister presence' behind the dynamic 'throne'.

The galactic nucleus in the centre of the system represents the third galaxy development motor. Previous chapters have already considered this from several angles. The galactic central area and the actual nucleus of our Milky Way system are not directly visible and not easily observed in other star systems either. In spite of this, observations show that the nucleus of the star system is not always so 'invisible', but is very much in evidence from time to time owing to explosions, gas ejection and extremely intense radiation in some spectral ranges and that it can influence the 'life' of the whole star system.

In 1943, the astronomer Carl K. Seyfert discovered a series of spiral systems which had distinctive, intense emission lines in their nuclear regions. These intense spectral lines can be explained by large gas clouds which are ejected from the centre at a rate of several hundred up to a 1000 km/s and are stimulated to give light. Other galaxies, investigated in great depth by the Armenian astrophysicist B. E. Markarian, show strong UV radiation in the central area, which is partly caused by hot stars and is partly synchrotron or magnetic bremssfrahlung. This radiation, too, is a pointer to special processes taking place in the central area. Explosions and 'jets' which are particularly associated with galaxies M 82 and M 87 and which can be seen direct on photographic plates are even more striking. These explosions, jets, increased UV radiation from the central area and emission line radiation are known as nuclear 'activities' and are associated with the ejection of more or less large quantities of gas matter and highly accelerated elementary particles. Although the nucleus of our star system is not so exciting at present, observations prove that it has lived through many 'stormy' periods in the past. Thus, some astronomers feel that the high-velocity clouds seen in high galactic latitudes, for example, are caused by explosive events in the galactic nucleus several million years ago and that the 135 km/s and 3 kpc arms were also probably caused by the same sort of nuclear activity. Hence the galactic nucleus is rightly assumed to be the third important factor affecting the evolution of the star system, even though we do not know how the 'motor' operates and what 'fuel' it uses.

In the Beginning There was Gas

The most important, but at the same time most difficult, aspect of astronomical research is to assemble the many individual facts into a corporate and united whole and thus present an image of the formation and evolution of the heavens, and of star systems in particular. However, a good deal of work is still based on assumptions and hypotheses which do have their origins in known physical laws but still remain unproven. In this sense, all modern concepts of the formation of the star system are hypothetical in nature, rather than cut and dried theories. A good deal of work is required before a more detailed plan can emerge.

According to one such hypothesis, about 15 billion years ago the world was in a state known as cosmological singularity and which was as follows: the whole of the tangible world was concentrated into a very small volume in which density was extremely high (however, this small 'world ball' could not be seen from the outside as there *was* nothing outside and no inside information could penetrate through). The matter contained in this world must have had very exotic properties, that is, assuming that we can even envisage individual properties under such circumstances. Hence we say amongst other things that matter at that time was in a state of 'proto-pulp'. At this stage of zero time, the 'world theatre' was opened up and the 'proto-pulp' began to explode and expand with incredible vehemence, hence the term 'big bang'. The first of our known elementary particles, predominantly neutrinos and antineutrinos, were formed in one ten thousandth of a second and after one hundredth of a second positrons and electrons emerged; these had a mutual exchange effect, i.e. high-energy light photons were formed from positrons and electrons and these

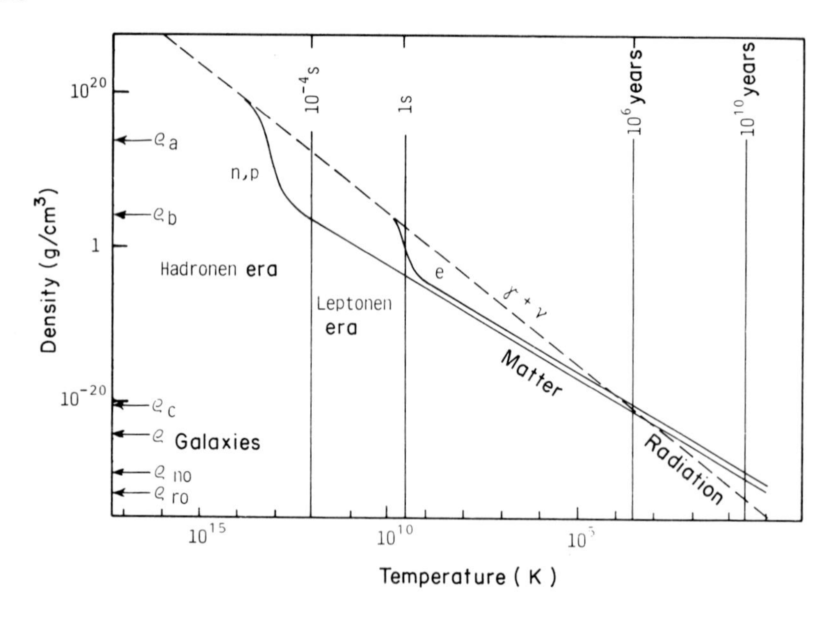

Figure X.1. Our world began to evolve about 15 billion years ago and was started off by the 'big bang'. Mostly heavy elementary particles (Hadronen) formed in the first ten thousandth of a second and started to interact: they gave out annihilation radiation and were re-formed. Only one in several billion of these particles survived this period. When the temperature dropped to 10^{12} K, the Leptonen era, in which electrons inter-reacted, began. After one second, the temperature had dropped so much that even electrons could no longer be formed by pairing. However, the radiation density was still greater than the material density and this did not change for a million years when photons were released from matter. Thus began the star era, which has continued to the present day. δ_a, δ_b and δ_c represent the material density at the end of the Hadronen, Leptonen and Radiation eras respectively, δ_{ro} and δ_{no} represent current radiation and material density, respectively, and $\delta_{galaxies}$ is the density of the larger galaxies. (After Harrison)

then gave rise to positrons and electrons once again. The electron–positron production processes were in equilibrium with the electron–positron destruction processes. Protons and neutrons were created and destroyed in a similar way. It is also assumed that matter and radiation were linked together. The rapidly changing 'proto-pulp world' expanded further and in doing so cooled down. During the period between one hundredth of a second and 100 seconds from zero time, protons combined with neutrons to form helium nuclei. Later, the particles had expended so much energy on expansion and subsequent cooling that there was no more left for building up any further elements. Thus, the Universe consisted essentially of protons, neutrinos, antineutrinos, protons and helium atoms and just enough electrons to balance out the positive charge of the protons and helium nuclei. At this point, radiation registered the same temperature as matter.

Some 300,000 years after the 'big bang', expansion was so far advanced that the free electrons recombined with the protons and the helium nuclei to form

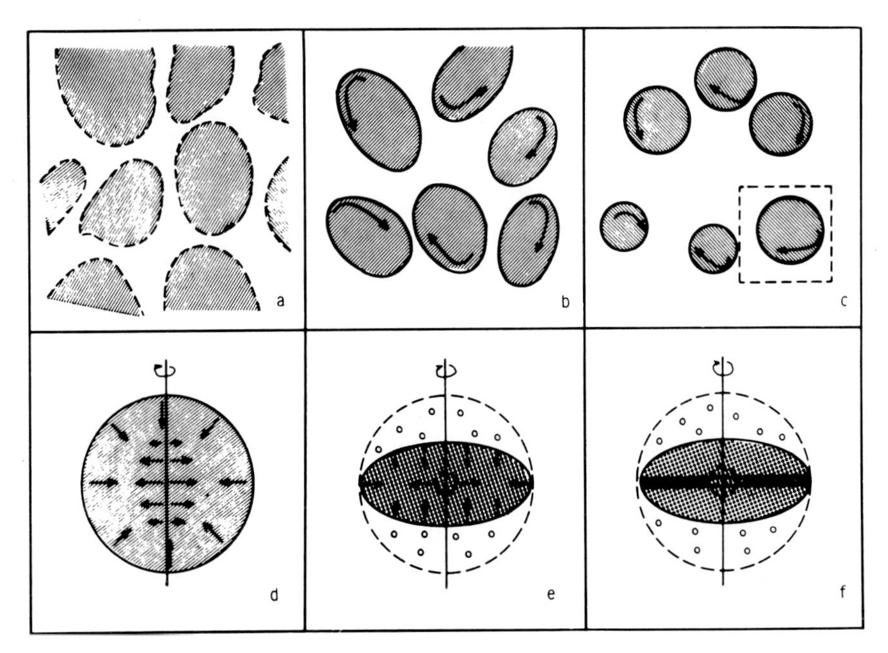

Figure X.2 About 300 million years after the 'big bang', the world was made up essentially of neutral gas. This gas was not uniformly distributed overall and there were more or less sizeable fluctuations in density (a). These began to contract and broke up into several cloud pieces which then continued to contract on their own (b). The individual sections were in fact galaxy forerunners (proto-galaxies). The first generation of stars, concentrated predominantly into globular star clusters, evolved within them (c). The residual gas contracted under the effect of gravity and rotation (centrifugal force) (d) and formed a disc in which further star generations evolved (e). The large and rapidly developing stars produced heavy elements by nuclear fusion and after their 'death' they bequeathed these to the surrounding interstellar medium. Instability in star and gas movement then produced the spiral pattern

neutral gas. From this point on, gas and radiation cooled down at different rates—radiation and matter were uncoupled. After 300 million years the world was filled with the original gas made up of hydrogen and helium (mixing ratio 9 : 1) and the residual radiation. With this gas the construction of the world as we now know it could begin. It contained incidentally many different levels of density, so that lower and higher density areas emerged. The high-density areas—'proto-clouds'—now began to pull in their own gravitational force. During the first phase of contraction, the 'proto-clouds' broke up into individual fragments and these continued to contract on their own. This cloud fraction phenomenon could have been an intermediate stage in the formation of the subsequent galactic clusters, e.g. local Group. As the fragments had also picked up a certain amount of rotary power along the way, further contraction started off cloud rotation, which in its turn ensured that the attractive force of the centre vertical to the axis of rotation was less than along the axis itself as a result of centrifugal force. This resulted in the flattening of

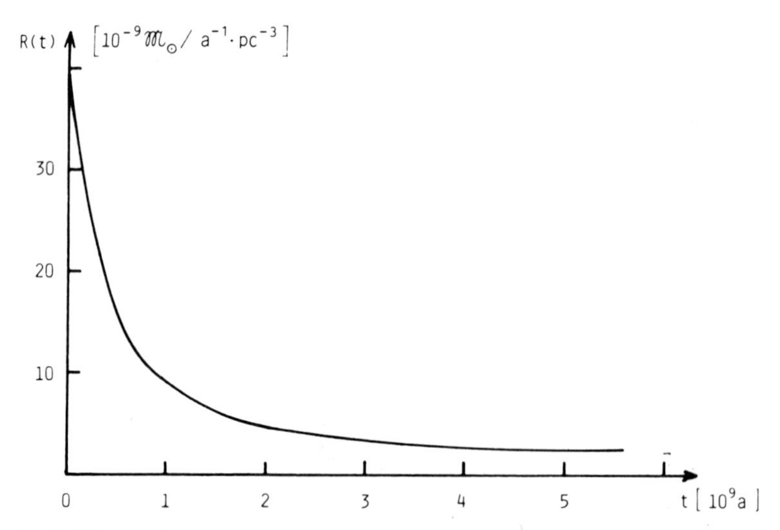

Figure X.3 When galaxy formation was in its early stages, the rate of star formation *R(t)* was much greater than it is today; 10–20 times as many stars were formed and more than half of all the stars were produced in the first few billion years. Since that time, the rate of star development has remained almost constant

the cloud and finally the formation of a disc with a thick centre section. Particles which revolved in the 'wrong' direction bumped into others, thus losing their rotary motion. However, the full force of gravity acted upon them and drew them into the centre, whilst direction and speed of rotation was decided by the 'majority'; hence the thickness of the central area. Before the disc was formed, some larger cloud sections had made themselves 'independent' and these then rotated around the 'mother cloud' in a more or less extended orbit. These formed the basis for the future globular star clusters.

Contraction and the ever increasing gas compression which followed continued until conditions were right for star formation. At the same time, the spiral structure probably began to build up in the disc. However, not all gas matter was converted to stars; about 10% was left over and this remained in the system as interstellar matter, keeping the process in action right up until the present time. On the other hand, star formation in the halo stopped as a result of a 'lack of raw material'. Because of this, the stars belonging to the halo population are 'first hour witnesses'. These stars are not made of material which has already been 'cooked through' and they have a relatively low metal content. Even their movement betrays the fact that they have been 'frozen out' of the general contraction pattern at a very early stage. They form the oldest objects seen in the Milky Way system.

Our system must have undergone a chemical change, during the course of which the relative amount of heavy elements (= metal content) increased whilst the overall element mixture in the star system remained remarkably uniform. Detailed analysis showed that the ratio of hydrogen to helium atoms is about 10 throughout and that the distribution of heavy elements is almost the same over the whole system. The ratio of the number of heavy atoms to that of hydrogen, or metal abundance, can be up to 500 times smaller in the halo population than in the Sun and in many disc population and spiral arm stars it is higher than in the Sun, e.g. about three times greater in the Hyades stars. In 1957 M. Burbidge and G. Burbidge, W. W. Fowler and F. Hoyle put forward a hypothesis in an attempt to clarify element distribution (known by the initials of the inventors as the B^2FH hypothesis). According to this, the primitive clouds were made up essentially of hydrogen and helium and, after star formation had been set in motion, higher elements were incorporated in the stars as they evolved further. Thus the matter discharged into interstellar space by the 'dying' stars had a much greater higher element content which it passed on to the next batch of stars being made. After just a few of these cycles, the element distribution as we know it today was established. However, this mechanism is possible only if the star formation and evolution in the early period had taken place very quickly, as even the older disc population stars were made up of almost the same mixture as they are now. However, only stars with a mass in excess of 5 sun masses could develop so rapidly, so most of the stars must have been massive at that time, quite different from today (the 'early galacticians' must have observed a different luminosity function to ours). As this type of luminosity function is not present even in the old stars of the halo population, some astronomers are having doubts about the theory at this point and suggest that maybe the heavy elements were 'cooked' in the galaxy nuclei and then moved as far as the outer areas of the system. Others favour the older hypothesis put forward by G. Lemaitre and G. Gamow, according to which the elements were already formed in essence during the 'big bang' inside the giant fireball. There is still no completely satisfactory answer to this question.

Explosions in the Centre

So far, we have seen how the halo and galactic disc sub-systems have been formed within the star system, how the spiral pattern emerged and the chemical changes which occur during star evolution. But what is happening in the centre? The high star density can be satisfactorily explained by star formation at high gas densities, as galactic contraction in the galaxy coincides with an increase in density in the centre. The presence of interstellar matter in the centre is quite different. What causes the rapidly expanding gas disc with its ring of molecule clouds and how are the expanding arms (135 km/s and 3 kpc arm) formed? One explanation puts it down to explosions in the centre, and nucleus activity observed in other star systems seems to support this. According to a model representation, there was a great explosion in our centre about 60 million years ago. At that time, interstellar matter which was not used up in star formation, was concentrated in a disc 1000 pc in diameter and 200 pc thick. The explosion took place in the centre of the disc, throwing out large quantities of hot gas in all directions. The hot gas discharged from the disc presumably left the star system and the gas forced into the disc compressed that

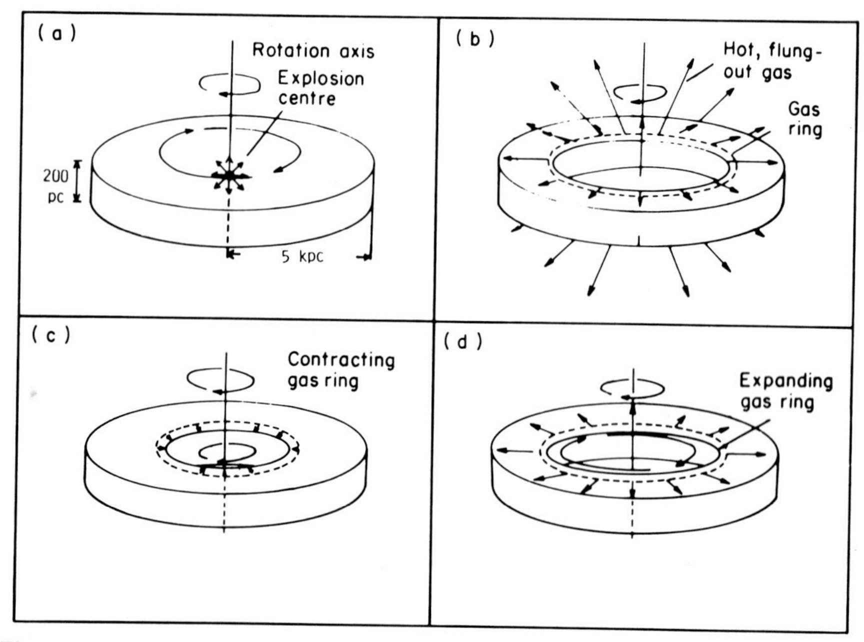

Figure X.4. According to one theory, the 3 kpc arm was formed as follows: a big explosion in the galactic centre heated the gas to a very high temperature and this then expanded in all directions (a). The expanded gas in the disc was compressed and thus formed an expanded, rotating ring about 3 kpc away from the centre (b). This ring continued to expand during the equilibrium gap between gravitation and centrifugal force, but gravity gradually gained the upper hand until expansion finally stopped and contraction began again. About 40 million years after the explosion, the ring had moved back to about 2.5 kpc from the centre (c). At this point, the centrifugal force was greater than the effect due to gravity, so the ring began to expand again. Some 60 million years after the explosion, the ring is again 3 kpc from the centre and moving outwards at a rate of 50 km/s (d) (After Sanders and Wrixon)

which was already there, forming an expanding, rotating gas ring about 3 kpc from the centre. The kinetic energy with which the gas was endowed during the explosion was so great that the ring expanded outwards beyond the point where gravitational and centrifugal forces were in equilibrium. However, during the course of time, the gravitational force became predominant and expansion stopped—indeed, the ring actually began to contract again. About 40 million years after the explosion, i.e. about 20 million years ago, the ring was pushed closer to the centre again, to about 2.5 kpc. The centrifugal force then took over once more (as the angular momentum, i.e. the product of mass times orbital speed and distance from the axis of rotation, which is a measure of rotational power, remains preserved overall, rotational speed is greater when distance is smaller; centrifugal acceleration is equal to orbital velocity divided by the distance from the axis of rotation). The 'game' thus began again—the ring expanded again until it reached its present position of 3 kpc and it is now moving outwards at a rate of 50 km/s. If corresponding calculations made by

R. H. Sanders and G. T. Wrixon do conform to reality, then the gas ring can continue to oscillate around this position of equilibrium for several hundred million years yet.

The molecule cloud ring can be considered in a similar way. This is 250–300 pc away from the centre and expands at a rate of 100 km/s. As its 50 km/s speed of rotation is too small to counteract the force of gravity, it is assumed that this ring was also ejected from the nuclear area (and molecule formation did not begin until after this event).

To create a gas ring such as the 3 kpc arm and its opposite number requires considerable power—in fact, an amount of energy corresponding to the complete conversion of more than 10,000 sun masses is needed [based on Einstein's equation: energy = mass × (velocity of light)^2]. Even the small molecule cloud ring requires 10 sun masses of energy. So what sort of object is it that catapults out so much gas and endows it with so much motional energy along the way? The theorists maintain that these requirements can only be satisfied by a supermassive star amounting to more than a million sun masses. Such a 'superstar' would be unstable and collapse under its own gravitational force, during the course of which gravitational energy would become converted to kinetic and thermal energy. One result of a release of thermal energy would be a continuous rise in temperature which could trigger off thermonuclear explosions. These 'giant hydrogen bombs' would then push out the gas clouds. On the other hand, if the conditions for thermonuclear explosions are not fulfilled, the superstar will collapse further, the density in the centre will become increasingly higher, the radius ever smaller and the acceleration due to gravity at the surface ever greater. When the latter becomes too big, even light can no longer escape. The superstar becomes invisible and we say that a high-mass 'Black Hole' has been formed.

Two mechanisms have been suggested as an explanation of superstar formation. Firstly, stellar collision could lead to great mass accumulation. However, for this mechanism to work, a star density of 3×10^9 stars pc^3 would be needed in the centre, which is 3000 times more than that actually calculated for the galactic centre. The second proposal is the so-called 'gas accretion' theory in which gas matter is collected together into one great body. As long as the gas has a high energy content, it moves rapidly and can escape the 'dust suction effect' of the gravitational forces in play within the growing body. However, in time the gas loses energy through radiation and can be drawn into the centre, providing favourable conditions for the emergence of a superstar.

Recent observations point to an upper limit of 2 million sun masses for a superstar in the centre of the star system, but no such object has actually been pinpointed. The upper mass limit implies that it has enough mass for one explosion at the most. Thus, if nuclear activity increased dramatically every 500 million years, as many experts think, then the gas must be supplied from outside the galactic nucleus area, during the course of which one tenth of a sun mass is required per year. Where does this gas come from?

The strange phenomena seen in the galactic nucleus pose questions to which

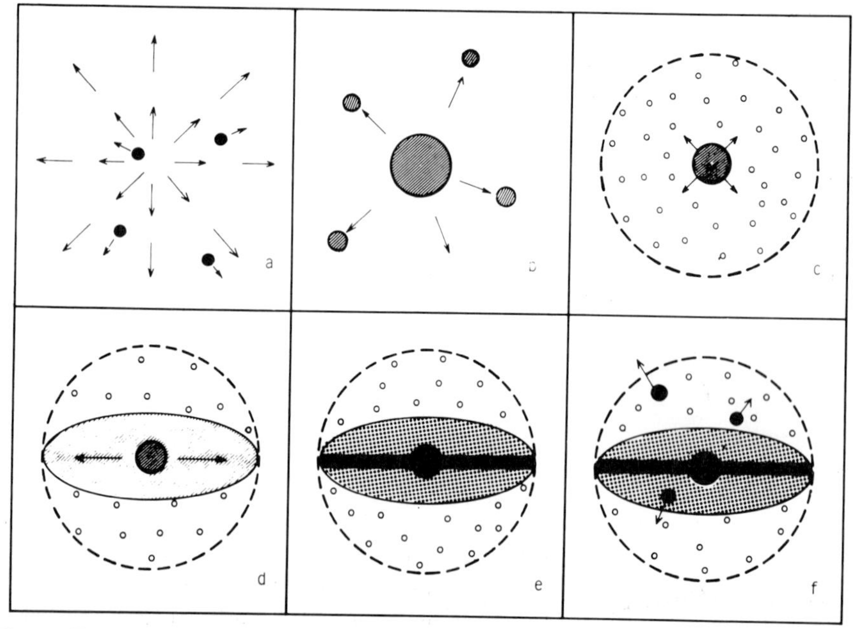

Figure X.5. According to one expansion theory of galactic formation, prestellar or pregalactic bodies were the main result of the 'big bang' (a). These were then further distributed (b) and the second generation pregalactic bodies represented the 'naked' galaxy nuclei. The various subsystems then 'gushed forth'. The first system to emerge was the globular cluster (c) and this was a result of nuclear fragments being flung out in all directions and small spherical star systems evolving around them. Then followed the disc and spiral populations, after which the galactic nucleus was virtually spent (d, e, f). Individual fragments of the nucleus which were flung further afield could be the basis on which companions such as the Magellanic Clouds were formed

we have no satisfactory answers. Much of this and the role played by the galactic nuclei during the evolution of the star system are based on hypothesis and speculation.

Maybe Everything was Quite Different?

In 1958, the Armenian astrophysicist V. A. Ambarzumjan gave a much acclaimed lecture at the Solvay congress in Brussels, in which he presented some basic ideas in the formation and evolution of the star system. He indicated that we still currently award the most important role in star formation to contraction and fragmentation mechanisms, but that expansion, explosion and activity are found in the nuclear areas of the star system on a much larger scale. Expansion processes are found in the gas envelope around Supernovae and Planetary Nebulae, in star associations and in many galactic clusters and star systems. Therefore, should not greater significance be attached to expansion processes, which are often involved in the various forms of nuclear activity, including that during the formation and evolution stages?

If we assume that contraction of the most varied forms of matter is not the basic process involved in the formation of heavenly objects, but that its opposite number, expansion, is, then the activity in the galactic nuclei becomes of central importance and the nuclei themselves become the most important objects in the formation and evolution of the star system. According to V. A. Ambarzumjan the nuclei set up galaxies around themselves during the course of this activity, i.e. they actually produce them. However, he did stress during his lecture that much more intensive observations were required in order to verify the whole concept at all levels. At the AGM of the International Astronomical Union in Berkeley, CA, USA, in 1961, he pursued his basic idea of the genetic role of the galactic nucleus. He emphasized that the originally 'naked' nucleus 'flings out' the galactic sub-systems as a result of various types of nuclear activity, and that even today the central origin is still obvious from the dense galactic concentration which surrounds it. Parts of the nucleus itself may have been catapulted out also under certain circumstances and each of these would then be capable of establishing its own small empire. Is this perhaps how the smaller galactic satellites, such as the Magellanic clouds belonging to the Milky Way, were formed? He went one step further and suggested that the special features of the spiral, such as associations, star clusters and giant H II areas, were born from 'nuclear fragments', and that the stars as a whole come from nuclei or nuclear sections.

Many observations point to the genetic nature of the galactic nucleus and to the significance of expansion processes. However, the above concept of evolution has in no way actually been proved by observation. We still have no idea what the dense prestellar material which forms the nucleus and which is capable of flinging out large amounts of matter and virtually ready-made stars looks like and how it conforms to the laws of physics as we understand them today.

The Galactic Nucleus as a Centrifugal Machine?

The explosion and expansion hypothesis would make the history of the Milky Way system roughly as follows: the world was started off with a 'big bang', which resulted not only gas and radiation emission but also in the formation of many large, massive, dense bodies made of prestellar (or pregalactic) matter plus existing radiation. Explosion and division then reshaped these bodies into nests of pregalactic elements which could have had different structures and represented the first stages in the evolution of galactic clusters. These pregalactic bodies were the 'naked' galactic nuclei which then set about 'dressing themselves with a star jacket'. Nuclear activity sent fragments of nucleus shooting out in all directions and each small section formed its own little colony, as a result of which the spherical sub-system or halo of globular clusters emerged. In phase two, a second sub-system, representing an intermediate population 'gushed' out of the nucleus and activity continued until the galactic disc finally emerged. Numerous mini-explosions could have catapulted stars and smaller star groups out of the nucleus. Alternatively, nuclear fragments and sections ejected into the plane may have continued to divide and develop, releasing stars, gas and dust, during the course of which these secondary activity centres could have made a very valuable contribution to the formation and development of the spiral structure. Even the star associations which are the main feature of the spiral arms and which often have particularly young star clusters in their central area may owe their existence to such activity centre. The Great and Small Magellanic Clouds could have been derived from a larger nuclear fraction flung rather further out and which also 'operated' as a secondary activity centre.

After the birth of the disc-shaped sub-system, the Milky Way nucleus was more or less exhausted; it became calmer and insignificant, although did not fade away completely and 'rumbled' from time to time, discharging various quantities of gas, e.g. the clouds forming the 3 kpc and the 135 km/s arm. However, the nucleus was basically fully spent after the creation of the Milky Way system, even though the relatively low activity of the galactic nucleus does continue to contribute to the further evolution of some areas, especially those near to the centre.

Although this hypothesis on the creation of the galaxy can clarify certain visual phenomena such as the ejection of gas clouds, galactic centre patterns, the associations as local structures and the Magellanic Clouds as companions to the Milky Way system, most modern astronomers favour the contraction theory. For one thing, the hypothesis is much more detailed and accurate as a result of an extensive programme of model calculations. The second point in its favour is that scientists are suspicious of the unknown prestellar or pregalactic matter, which is excessively dense and has highly exotic properties. The third and final reason is that Ambarzumjan's hypothesis has not yet been adequately substantiated. Indeed, he has frequently stressed the fact that more material based on actual observation of the various forms of nuclear activity in particular must be available before any concrete theories can be drawn up. Hence there can be no decision as to the winner of the contest for the best theory of the creation of the star system, although the balance seems to be weighted more towards the contraction hypothesis for the present.

XI

Why astronomers are interested in the star XYZ

Our 'illustrated and explanatory walk' through the Milky Way is in fact over, and this little book could now be shut. However, there are still a few queries which need to be clarified: what is it all in aid of? The large amounts of time and money spent on astronomical research cannot be simply for the benefit of a group of intellectuals. What does it matter whether another system, several million parsec away and not even visible to the naked eye, exploded or not, or whether a star seen through the largest telescope is 20 million or 5 billion years old? The average man in the street would regard the whole problem as insignificant and would experience neither joy nor sorrow on learning the correct answers. However, this is what everyday research work is all about and much large-scale equipment is required to do it, including enormous telescopes and gigantic radio telescopes, plus expensive electronic equipment and computers. Astronomical telescopes with all their accessories are extremely costly—in fact they and the accelerators used in high-energy physics represent the most costly items of all scientific research equipment. We should therefore take a good look at the whys and wherefores of astronomical research.

Our first point comes under the heading 'astronomers produce a scientific view of the world'. Obviously they are not the only ones involved, but join together with philosophers, social scientists and other scientists such as biologists, chemists and physicists towards a common goal. Astronomers have a very important role to play as they actually 'construct' the framework in which our small world which is so valuable to us is 'suspended'. Astronomical findings based on physical laws provide the background for the solution to questions such as, 'Where is our Earth?', 'What does the overall world pattern look like?', 'Under what conditions did the world evolve?' and 'Where do we go from here?'. As astronomers find the answers to these questions, they provide accurate information on the material make-up of our world, but a vital intermediate stage inevitably involves working on many other, often banal aspects (whether a star is 20 million or 5 billion years old is a typical query which falls into this category) and the answers as a whole then reproduce the great mosaic which forms the astronomical concept of the world.

The outstanding role played by astronomy in establishing the modern scientific view of the world is often referred to in this context and its current and future existence is justified on this basis. It is a fact that modern scientific concepts are closely connected with and indeed would not exist without the Copernican revolution and such names as Galileo, Kepler and Newton, who were all deeply involved in astronomy. However, that is a historical argument and represents a very minor contribution to the solution of the main question posed by this chapter. The Copernican revolution was the vital major step in establishing the correct scientific view of the world, but this sort of step is unique and since that time all astronomical findings have filled in a few details, but brought about no fundamental changes. That is not to dismiss the work of many astronomers, but shows that astronomy cannot rest on its laurels, neither can it justify its philosphical existence by waiting for a new Copernican-type revolution. Nowadays, astronomy is of no greater importance than any other scientific discipline.

Let us now return to our original question and try to demonstrate a second equally philosphical point of view. Our modern techno-commercial world is based on a scientific concept of the world without which there would be no such things as calculators, television, aeroplanes and atomic power. Astronomy has contributed to this image and indeed continues to do so. Therefore, we could maintain that now that astronomy has played its part out and we know all about the world both around us and as a whole, that is enough, the observatories should be shut and the money spent on much more useful things. However, that would be rather a short-sighted conclusion, for scientific knowledge in general is not a specific once and for all affair to be stored in the brain and in libraries for all time. It is much more an on-going process demanding a good deal of continual effort. It can be compared to a man swimming against the current in a river; when he stops trying, not only does he make no further process, but he is actually dragged back by the force of the water. Research is just like that; as soon as questions stop, problems are no longer tackled and details abandoned, and not only does nothing new emerge but the validity of previous knowledge gained and its correct application are quickly forgotten. Then hypotheses could easily become dogmas and some existing knowledge applied to uncertain areas and methods by means of which this same information was acquired could be totally forgotten. To make this all sound convincing is a difficult task, but there are historical examples which clarify the point. Greek materialists and physicists (in so far as we can apply these modern titles to such scholars as Archimedes and Heron) had already reached a considerable level of advance by modern standards (the list of names could obviously be extended to cover many other Greek and Roman philosophers), but this lay forgotten by the whole of Europe for a very long time and during the revival period known as the Renaissance scientists had to go right back to basics and start again from scratch. We have already established that astronomy makes a valuable contribution to the realm of scientific knowledge; therefore, it cannot be put on ice and excluded from the general process of learning. Astronomy also plays its part in ensuring world progress.

Perhaps it is a good thing to examine the question 'Why is so much money allocated to astronomy?' from as many aspects as possible, incorporating a few other important arguments in favour of astronomical research at the same time, but without letting it become the most burning topic of the day. An apt heading for this section would be 'Scientific research is a cultural achievement'.

In addition to the absolutely essential requirements for life, such as eating, drinking and sleeping, there are a whole series of less urgent needs, e.g. a nicely established home with more than just a table, cupboard and bed in it, and a suitable amount of leisure time in which to enjoy it. We often lump all these things together under the general term 'culture'. The human facility for recognizing and reflecting the environment falls into this category too. It can be viewed from many angles and art in its many and varied forms is certainly one of the more important. Scientific research helps us to look at the environment objectively by isolating the general interrelationship patterns of nature all around us. These laws then form the basis of a better understanding of nature, which in its turn leads to heightened knowledge of mankind itself. This in fact makes the point that the cultural level of a country is very dependent on its general state of scientific advance. Obviously, astronomy also contributes towards the general knowledge bank and to exclude it from the whole would be unthinkable. At the same time, any branch of science must be for the people. Astronomy contributes to our cultural standards; apart from navigation and time measurement, it helps to increase our awareness of the world around us and thus of ouselves and, the most important factor of all, it helps us to appreciate our knowledge.

The very useful by-products of astronomy should be mentioned as a follow up to the 'astronomy for the people' concept. There are a whole range of astronomical discoveries and methods which have been used by physicists for very practical purposes. The list starts with celestial mechanics, including all its laws and methods, which forms the basis of space travel and extends to the hydromagnetic properties of ionized gases which were first investigated in space and are now an integral part of the structure of hydromagnetic generators for the production of electricity and controlled nuclear fusion. The LASER is an example of this; it was not built or even conceived by astrophysicists, but it certainly has its theoretical basis in astronomical research.

Research into the spectra of bright gas nebulae in interstellar space revealed clear spectral lines which actually should not have been there. There was no known element to which they could be ascribed, hence 'nebulium' was discovered. These lines were found in the oxygen atom for certain quantum figure combinations in the field of quantum mechanics. However, in accordance with the well established prevailing laws, these combinations were forbidden and were therefore known as forbidden lines. Strictly, they are not forbidden, of course; the corresponding transition stages between the various atomic energy levels have a very small probability factor (e.g. a hydrogen atom which wants to emit on a 21 cm wave—that is a forbidden line too—will have to wait in an excited state for about 10 million years before this happens spontaneously, neither must it be substantially distributed during this time). However, in areas of high gas density (which are present even in a laboratory vacuum) the atoms do not remain undisturbed for long enough, there is too much collision and so the lines are not seen. Only astronomy was able to show that they do actually occur under conditions prevailing in space. LASERs operate in the forbidden line range and the forbidden line theory is an integral part of their theoretical concept.

These sorts of illustrations point to the conclusion that space is a 'laboratory' with physical conditions which cannot be simulated on Earth as yet, and that physicists are grateful for the existence of this laboratory. In this way, astronomy, astrophysics and physics itself often go hand in hand.

Astronomical research has yet another useful aspect to it; the equipment used is very costly and, because there is so much money involved, the manufacturer must be highly aware of developments in astronomy in order to construct new and better instruments. This involves close collaboration between astronomers and manufacturers, and ensures a high level of progress in the machine industry.

We have seen that it makes no sense to rationalise out because of its subject so far from Earth. However, it still remains an 'expensive pastime' and most of the money must come from the general public. Therefore, the general public has a right to know what the money is spent on and hence astronomers have a duty to make the results of their research both available to and understandable by the average man in the street. This stipulation will ensure that as many people as possible can take part in the cultural benefits of scientific research, either directly or indirectly. Albert Einstein once wrote, 'It is most important that the general public should be given the opportunity of learning about and participating in the objectives and results of scientific research. To confine each result to a few specialists in that particular field is not enough. If the whole of knowledge is restricted to a small group, then all philosophical thought will die and spiritual poverty will reign supreme'. This is also the objective I had in mind whilst writing this book and I resisted the temptation of friends to join their merrymaking activities on many a long evening until my goal was achieved.

Appendix

Table 1. Units of measurement

Distance	1 astronomical unit (AU)	$= 1.496 \times 10^{11}$ m
	1 parsec (pc)	$= 2.063 \times 10^{5}$ AU
		$= 3.08572 \times 10^{16}$ m
	1 kiloparsec (kpc)	= 1000 pc
	1 sun radius (R)	= 695,990,000 m
Mass	1 sun mass ($M_{\odot}$)	$= 1.989 \times 10^{30}$ kg
Density	1 hydrogen atom/cm^3	$= 1.67343 \times 10^{-24}$ g/cm^3
	1 $M_{\odot}$/pc^3	$= 6.770 \times 10^{-23}$g/cm^3
		= 40.5 hydrogen atoms/cm^3
Energy	1 erg (cm^2 g s^{-2})	$= 2.7 \times 10^{-14}$ kWh
Luminosity (output)	1 sun luminosity ($L_{\odot}$)	$= 3.83 \times 10^{33}$ erg/s
		$= 3.83 \times 10^{23}$ kW
Wavelength	1 Ångstrom (Å)	$= 10^{-10}$ m
	1 nanometer (nm)	= 10 Å = 10^{-9} m

Table 2. Molecules observed in interstellar space

Type	Inorganic		Organic	
Diatomic	H_2	Hydrogen	CO	Carbon monoxide
	OH	Hydroxyl	CN	Cyanide
	SO	Sulphur monoxide	CS	Carbon monosulphide
	SiO	Silicon monoxide	CH	
	NS	Nitrogen sulphide	CH^+	
	SiS	Silicon sulphide		
Triatomic	H_2O	water	CCH	Ethynal
	H_2S	Hydrogen sulphide	HCN	Hydrogen cyanide
	SO_2	Sulphur dioxide	HNC	Hydrogen isocyanide
	N_2H^+		HCO	
			HCO^+	
			OCS	Carbonyl sulphide
Tetratomic	NH_3	Ammonia	$CCCN$	
			H_2CS	Thioformaldehyde
			H_2CO	Formaldehyde
			$HNCO$	
Pentatomic			H_2CNH	
			NH_2CN	Cyanamide
			H_2CCO	
			$HCOOH$	Formic acid
			$HCCCN$	Cyanoacetylene
Hexatomic			CH_3OH	Methanol
			CH_3CN	Cyanomethane (acetonitrile)
			$HCONH_2$	Formamide
Heptatomic			$HCCCCCN$	
			CH_3CCH	Methylacetylene
			CH_3CHO	Acetaldehyde
			H_2CCHCN	Vinyl cyanide
			CH_3NH_2	
Octatomic			$HCOOCH_3$	
Nonatomic			$(CH_3)_2O$	Dimethyl ether
			C_2H_5OH	Ethanol
			C_2H_5CN	
			$HCCCCCCCN$	

Table 3. Stars not more than 5 pc distant

Star	m_v	M_v	$L/L_\odot$	r	Sp
Sun	−26.8	4.8	1.0	8	G 2
α Centauri A	0.0	4.5	1.0	1.32	G 2
α Centauri B	1.4	5.9	0.36	1.32	K 6
α Centauri C	11.0	15.4	0.00006	1.32	M 5
Barnard's star	9.5	13.2	0.00044	1.81	M 5
Wolf 359	13.5	16.7	0.00002	2.32	M 8
BD 36° 2147	7.5	10.5	0.0052	2.49	M 2
Sirius A	−1.5	1.4	23.0	2.65	A 1
Sirius B	8.3	11.2	0.0028	2.65	DA
Luyten 726–8A	12.5	15.3	0.00006	2.74	M 6
Luyten 726–8B	13.0	15.8	0.00005	2.74	M 6
Ross 154	10.6	13.3	0.00004	2.90	M 6
Ross 248	12.2	14.7	0.00011	3.15	M 6
ε Eridani	3.7	6.1	0.30	3.28	K 2
Luyten 789–6	12.2	14.6	0.00012	3.31	M 6
Ross 128	11.1	13.5	0.00033	3.32	M 5
61 Cygni A	5.2	7.5	0.083	3.42	K 5
61 Cygni B	6.0	8.3	0.04	3.42	K 7
ε Indi	4.7	7.0	0.13	3.44	K 3
Procyon A	0.3	2.6	7.6	3.48	F 5
Procyon B	10.8	13.1	0.0005	3.48	
Σ 2398 A	8.9	11.2	0.0028	3.52	M 4
Σ 2398 B	9.7	12.0	0.013	3.52	M 5
BD 43° 44 A	8.1	10.4	0.0058	3.55	M 1
BD 43° 44 B	11.0	13.1	0.0004	3.55	M 6
CD 36° 15693	7.4	9.6	0.012	3.58	M 2
τ Ceti	3.5	5.7	0.44	3.66	G 8
CD 39° 14192	6.7	8.8	0.025	3.85	M 1
Kapteyn's star	8.8	10.8	0.084	3.91	M 0
Krueger 60 A	9.7	11.7	0.0017	3.94	M 4
Krueger 60 B	11.2	13.2	0.00044	3.94	M 6
40 Eridani A	4.4	6.0	0.33	4.88	M 5
40 Eridani B	9.5	11.2	0.0027	4.88	DA
40 Eridani C	11.2	12.8	0.00063	4.88	M 4

m_v = apparent visual brightness
M_v = absolute visual brightness
L = luminosity
r = distance in parsec
Sp = spectral type

Table 4. The 20 brightest stars in the sky

Star	m_v	M_v	$L/L_\odot$	r	Sp
Sirius	−1.47	1.4	23	2.7	AIV
Canopus	−0.73	+3.0	130	56	FOIb
α Centauri	−0.01	4.5	1.1	1.33	G2V
Vega	0.04	0.5	52	8.1	AOV
Arcturus	0.06	−0.2	100	11.1	K2III
Rigel	0.08	−7	52,000	250	B81a
Capella	0.09	−0.6	145	13.7	G8III ± F
Procyon	0.34	2.6	7.6	3.48	F5
Archemar	0.47	−2.7	1000	44	B5IV
β Centauri	0.59	−3.4	1900	62	BIII
Altair	0.77	2.3	10	5.1	A7V
Betelgeuse	0.80	−5	8300	150	M21ab
Aldebaran	0.86	−0.7	160	20.8	K5III
Spica	0.96	−2.4	760	48	BIV
Antares	1.08	−2.5	830	53	MIIb
Pollux	1.15	1.0	33	10.8	KOIII
Fomalhaut	1.16	2.0	13	6.9	A3V
β Crucis	1.24	−5	8300	176	BO.51V
Deneb	1.26	−7	52,000	450	A21a
Regulus	1.36	−0.7	160	25.8	B7V

m_v = apparent visual brightness
M_v = absolute visual brightness
L = luminosity
r = distance in parsec
Sp = spectral type

Table 5. Milky Way data

Diameter of galactic disc	30 kpc
Thickness of galactic disc	1 kpc
Diameter of central area	5 kpc
Diameter of galactic halo	50 kpc
Mass of Milky Way system	$1.2–1.9 \times 10^{11}\ M_\odot$
Average density	$0.1\ M_\odot/pc^3$
Density in the Sun's vicinity	$0.1\ M_\odot/pc^3$
Density in the centre: 1 p-c from centre	$400{,}000\ M_\odot/pc^3$
10 pc from centre	$7000\ M_\odot/pc^3$
100 pc from centre	$100\ M_\odot/pc^3$
Total mass of the 'central star cluster'	$1.1 \times 10^{10}\ M_\odot$
Thickness of galactic gas layer	250 pc
Mass component of interstellar gas	10%
Mass component of interstellar dust	0.1%
Number of interstellar clouds in the radius vector	7–10 per kpc
Number of associations and open star clusters (estimated)	10,000–100,000
Number of globular clusters (estimated)	2000
Distance of Sun from centre	10 kpc*
Distance of Sun from galactic plane	14 pc
Rotation speed in the vicinity of the Sun	250 km/s
Revolution period of Sun	250×10^6 years
Epicycle revolution period in the vicinity of the Sun (epicycle frequency)	200×10^6 years
Period of vibration vertical to the galactic plane (T_z)	70×10^6 years
Rotation speed of spiral pattern (Sun's vicinity)	125 km/s
Spiral arm diameter	~1 kpc
Spiral arm distance	~2 kpc
Gradient angle of spiral arms	~15°
Age of Milky Way	$1.1–1.5 \times 10^{10}$ years
Strength of galactic magnetic field	2×10^{-6} Gauss

(continued)

Table 5. Milky Way data (*continued*)

Populations		
Spiral arm population (Pop. I): $\bar{z}$		0.12 kpc
$\bar{v}_z$		8 km/s
Members		Interstallar matter
		Associations
		Young star clusters
		OB stars
Disc population: $\bar{z}$		0.4 kpc
$\bar{v}_z$		17 km/s
Age		2–12 × 10^9 years
Members		Galactic disc and central area stars
Halo population (Pop. II): $\bar{z}$		2.0 kpc
$\bar{v}_z$		75 km/s
Age		12–15 × 10^9 years
Members		Globular clusters RR-Lyrae stars with Cycles greater than 0.4 days†
Interstellar matter in the centre		
3 kpc arm	Distance from centre	3–4 kpc
	Radial velocity	53 km/s
	Mass	$10^7 M_\odot$

135 km/s arm (expanded arm)	Distance from centre	2.4 kpc
	Radial velocity	135 km/s
	Rotation speed	170 km/s
	Mass	$10^7 M_\odot$
H I disc	Distance from centre	0–0.8 kpc
	Thickness (in centre)	80 pc
	Thickness (outside)	250 pc
	Maximum rotation speed	230 km/s
	Mass (H_2)	$10^7 M_\odot$
	(H I)	$4 \times 10^6 M_\odot$
	(dust)	$10^5 M_\odot$
Molecule cloud ring	Distance from centre	0.3 kpc
	Radial velocity	130 km/s
	Rotation speed	50 km/s
	Thickness	70 pc
	Mass	$10^8 M_\odot$
H II region in centre	Distance from centre	0.0–0.2 kpc
	Mass	$10^6 M_\odot$

*More recent values are between 8.5 and 9.5 kpc; as Milky Way models have always been based on a figure of 10 kpc, we shall continue to use this for the present.
†$\bar{z}$ = average distance from the galactic plane.
$\bar{v}_z$ = average speed vertical to the galactic plane.

Table 6. Members of local groups

Name	Type	r	m_B	M_B	D	Mass
Milky way System	Sbc	–	–	−19	36	2×10^{11}
Great Magellanic Cloud	SBc	0.052	0.63	−18.1	11	1.4×10^{10}
Small Magellanic Cloud	Ir	0.063	2.8	−1.0	4.6	
M 31 (Andromeda Nebula)	Sb	0.69	4.4	−20.3	40	3.7×10^{11}
M 32	E 2	0.69	9.1	−15.6	2.4	4×10^{9}
NGC 205	E 6	0.69	8.9	−15.8	5.2	
M 33 (Triangulum Nebula)	Sc	0.72	6.3	−18.3	18	1.4×10^{10}
NGC 147	E 4	0.69	10.6	−14.4	3.6	
NGC 185	E 0	0.69	10.3	−14.7	2.8	
NGC 6822	Ir	0.50	9.3	−14.8	2.9	
IC 1613	Ir	0.66	10.1	−14.2	4.2	
Wolf–Lundmark System	E 5	0.87	11.2	−13.7	3.3	
Leo System 1	E 4	0.23	11.3	−10.7	0.8	
Leo System 11	E 1	0.23	12.9	−9.1	0.7	
Sculptor System	E	0.11	9.2	−11.2	0.7	
Fornax System	E	0.23	9.1	−12.9	4.0	
Draco System	E	0.06			0.23	
Ursa Minor System	E	0.08			0.3	

r = distance from Milky Way centre in megaparsec (1 mpc = 10^6 pc).
m_B = apparent brightness.
M_B = absolute brightness.
D = diameter in kpc.
Mass is given in sun masses, $M_\odot$

Index

Page numbers in *italic* refer to figure illustrations.